The Alarm Dealer's Guide

The Alarm Dealer's Guide

John Sanger

Illustrated by Jay Watson Jones

BUTTERWORTH PUBLISHERS
Boston • London
Sydney • Wellington • Durban • Toronto

All references in this book to personnel of male gender are used for convenience only and shall be regarded as including both males and females.

Library of Congress Cataloging in Publication Data

Sanger, John.
The Alarm Dealer's Guide.

Includes index.
1. Burglar alarm industry—Management.
2. New business enterprises—Management. I. Title.
HD9999.S452S26 1985 621.389′2 84–17484
ISBN 0–409–95088–2

Butterworth Publishers
80 Montvale Avenue
Stoneham, MA 02180

10 9 8 7 6 5 4 3 2 1

Printed in the United States of America

To my wife, Sandy,
to my family,
and to Mary (Grandmaw Liberty) Sager,
for their support and encouragement.

"Now go, write it before them in a table,
and note it in a book."
Isaiah 30:8

Contents

PART II. TECHNIQUES

PART III. APPLICATIONS

PART V. SOURCES

Preface

"Cursed be he that removeth his neighbour's landmark."
Deuteronomy 27:17

The security industry in general and the alarm industry in particular are experiencing rapid growth. Innovative technologic advances and a new awareness of the market's potential have combined to thrust our industry into the forefront of American commerce.

The average person is becoming security literate, that is, he has gone beyond an awareness of security needs; he is becoming knowledgeable about loss-prevention methods as well as about security devices.

With increased security literacy in the public comes an increased responsibility for professionalism from those of us in the industry. It is a responsibility that we should welcome.

Security literacy is being effected from various sources. The media, in addition to reporting news of crimes and losses, are now presenting special programs on crime prevention. From local features to national television coverage, the public is receiving more and better information. Civic groups are using local security professionals to educate their members and to spread the word throughout the community. Large and small law enforcement agencies are assigning personnel to or increasing the staffs of crime prevention units. Institutions of higher education are offering courses designed for the public as well as for the security professional; courses ranging from home security to executive protection may now be found on college campuses.

The opportunity to take an active part in the community and at the same time contribute to the professional growth of the industry has never been greater. Ours is a vibrant, dynamic profession that is characterized by industrious and creative men and women. Our success depends, at least in part, on continued professional growth.

This book is written for the alarm dealer and installer. Its main purpose is to guide. There are no magic formulas, and the final measure of success is in the net profit figure on the income statement. There are, however, some basic guidelines for operating the business, obtaining customers, estimating and bidding jobs, as well as for planning and installing alarm systems that will help to ensure success—and profit.

As a guide, this book leads readers through all the necessary segments of business operations—from selling to installing electronic alarm systems. The book's approach is simple. It starts with a look at the market for security products and services and helps readers to identify niches for their businesses. Once the market has been defined and analyzed, it conducts the readers into general, easy-to-understand discussions and analyses of starting, organizing, and managing their businesses.

Specific and practical information on operating an alarm company is also presented. Daily accounting and inventory procedures are outlined, and other topics, like credit and collections and marketing and advertising, are explored in detail with particular emphasis on those areas unique to alarm company operations.

Numerous techniques for selling security products and services are included. A step-by-step approach to making a sales call, designing a system, bidding a job, and closing the sale is presented, which rounds out the first two sections of the book, "Strategies" and "Techniques." The remaining sections address the technical aspects of operating an alarm installing company: equipment installation. The applications section begins with an overview of basic electronics that will serve as a learning tool for novice installers and a refresher course for veterans.

Once readers have achieved a basic understanding of security electronics, they are presented with detailed explanations of each type of component used in intrusion alarm systems. Also included are numerous installation and application hints.

There are a couple of things to keep in mind. First, any time someone promises you a rose garden, remember that there are thorns among the beautiful flowers. Second, when thinking about success, consider a statement written by George Bernard Shaw in *Mrs. Warren's Profession:* "People are always blaming their circumstances for what they are. I don't believe in circumstances. The people who get on in this world are the people who get up and look for the circumstances they want, and, if they can't find them, make them."

On that note, let me wish you success in the alarm industry.

John Sanger

Acknowledgments

Special thanks are extended to Susan A. Whitehurst, editor, and the entire staff of *Security Distributing & Marketing* for their assistance with this book. A very special "thank you" goes to a person I admire and respect: Jerry Johnson, who is now retired from Adcor Electronics, Inc. Thanks, Jerry, for being a friend.

Many people provided information while this book was being written. They include: Norval ("Bud") Poulson (Alarm Industry Research and Education Foundation), Richard M. Bugbee (National Burglar and Fire Alarm Association), Dick Mellard (National Crime Prevention Institute), Bill Cook (Always Alert), Steve Heritage (Acron Corp.), Howard Berger (Alarm Controls Corp.), Bob Miller (Blue Grass Electronics, Inc.), Tom Scoggins (Diversified Marketing & Manufacturing Co.), Dante Monteverde and Maureen Gold (Emergency 24), Bob Rankin (Federal Signal, Autocall Div.), Harry Crawford (Foilmaster, Inc.), Max Riordan (Maxwell Alarm Screen Mfg. Co.), Don Johnston (Microwave Sensors, Inc.), Ken Rosenberg (Napco Security Systems, Inc.), Jeanne McGuire (PULNiX America, Inc.), Jim Murray (Raytek, Inc.), Ron Davis (Security Alliance Corp.), Ira Zipkin (Security Fabrics, Inc.), Charles Huckins (Sentrol, Inc.), Don Bosak and Larry Fiore (Transcience, Inc.), and many others.

My sincere thanks to all.

The Alarm Dealer's Guide

Introduction

It is hoped that this book will not be placed on the bookshelf and forgotten. It will not be a valuable tool if it is not used. It should be used frequently and it always should be readily available.

I encourage you to read this book with a pencil or marker in hand: underline, circle, check, and highlight. Make notes in the margins. Jot down ideas generated from reading a passage.

Remember, this is a working book. One of the best ways to absorb the information is to read it through once, leisurely and in its entirety, and become familiar with it. Then read it again and make notes.

While reading, compare the information with your own business and your specific job within the organization. If you are a small-business owner, you probably are wearing the hats of several people. It is unlikely that you do all of the things exactly as they are presented. Some may be similar; some completely different.

Maybe some of the concepts that are presented can be used without modification, maybe they can be used with a few changes, or maybe they cannot be used at all. The important thing is that you have made a comparison; you have measured your methods against those of others. Even if there are no changes made in your operations as a result of the comparison, you can feel more confident. Possibly, new ideas will be generated that can be used to increase profits.

There are few "absolutes" in the alarm industry, and those are usually technical (like Ohm's law) or financial (like the bottom line of a profit-and-loss statement). Most of what is encountered in the daily business operations is somewhere in the gray area between black and white.

For some time, I have suspected that if six alarm dealers or installers were called upon to design an alarm system for a structure, six different systems would be proposed. Differences would exist in the type, quantity, and location of equipment, as well as personal preferences for specific brands of equipment. Assuming that the six system designers were security professionals, it is doubtful that any of the proposals would be wrong. Instead, there would be six solutions for a single problem, any one of which would achieve the desired results.

On the one hand, those of us who are dealers/installers understand how so many different proposals could be presented. Each dealer has certain types of equipment that he prefers for a variety of reasons. Installation techniques and abilities differ. The dealer's background and experience also play a major role in designing a system.

On the other hand, the person receiving the different proposals probably will not understand. Therefore, the dealer should be able to point out the advantages of his system over those of his competitors.

If this book spawns just one idea that increases efficiency, reduces costs, or promotes a sale, it will have more than paid for itself. Because business operations change, rereading the book periodically may generate other ideas. Techniques that did not apply yesterday may apply today; and, if not today, maybe tomorrow.

The vast majority of alarm companies are small businesses, and this book is written with them in mind. Large alarm companies will find it useful, too, because many of their functions are similar to those of small businesses—they are just done on a larger scale and more frequently.

CHARTING A COURSE

The Alarm Dealer's Guide helps you chart a course to enter, proceed, or change direction in the alarm industry. It is designed to provide you with a logical and easy-to-follow pathway.

The embarkation point for any person planning to enter a new industry is a market analysis. For industry veterans, it also serves as a starting point for re-evaluating market positions and launching new marketing programs. That is why a look at the alarm market is included in Part I, "Strategies," as Chapter 1.

Continuing in a methodical manner, "Strategies" helps you plan your market entry by guiding you through the steps necessary to set up an alarm installing business. This part aids you in determining capital requirements and organizational structure.

Also included in Part I are guidelines for managing the organization, presenting a company image, and operating the business. Accounting, inventory control, supplier selection, credit and collections, and personnel policies are a few of the topics covered, addressing the unique requirements of alarm companies.

The first two logical steps, analyzing the market and establishing a business organization, are followed by the most critical part in any business enterprise: marketing goods and services. Part II, "Techniques," helps you do just that. It begins with an analysis of advertising media and ends with closing the sale. In between, you will find information on making sales calls, designing systems and bidding jobs. It's all presented with one goal in mind: to help make your alarm installing company successful.

The technical side of the alarm industry is discussed in Part III, "Applications." If you have been successful in marketing your goods and services, you have reached the test of your company's mettle: installing an alarm system.

From basic electronic procedures to specialized equipment and unique applications, Part III helps lead you through the maze of new and sophisticated alarm equipment. You will discover what the equipment does and how you can use it to provide your customers with security that meets their budgets and lifestyles.

Part IV, "Potpourri," addresses several topics: computerization, legal liabilities, false alarms, and licensing. Each can have significant effects on your business.

The days of a noncomputerized office are fading. Computerization may be the answer to becoming and/or remaining competitive and providing the services your customers require.

Your responsibility, and legal liability, to your customers is increasing—so are litigations in which alarm dealers are involved.

The most pressing issue for the alarm industry is false alarms. It is a problem that is affecting the professional credibility of the industry. The final chapter in this book addresses this problem.

Part V, "Sources," will help you throughout the book. It consists of appendices that contain information to help you do your job successfully.

We have come full circle: from market analysis through establishing and operating a business to system installation and current issues. Now it's time to start over.

You will need to follow the same logical approach each time you enter a new market or expand an old one. You will need to review the sections on organizing and operating a business when you open branch offices and when your organization structure requires changing.

Your business is ever-changing and dynamic—not static. It requires constant attention if it is to be profitable. This book can guide you down the road to profitability.

STRATEGIES I

Entering the Alarm Market 1

Today's business environment, especially the security/alarm industry, is characterized by change. The movement toward security literacy is changing the composition of the market, and alarm dealers must be aware of what is going on and remain flexible in their planning.

MARKETING AND THE MARKET

The words market and marketing have different meanings. *Marketing* is a broad term covering the entire process of moving goods (and services) from the manufacturer to the end user. *Market,* however, refers to the combined demand of potential buyers for a product or service. Markets may be very broad or very specific. Typically, the small alarm dealer is interested in a rather narrow market, such as a small town or an area of a larger town. Depending on the goals established for his business, he may have an even narrower market. For example, he may be seeking a market that consists of homeowners whose homes are valued in excess of $100,000, or only small retail establishments. In any case, the dealer should know *exactly* what his market is.

Market Research

Market research, whether performed by the dealer's staff or by professional researchers, is primarily a fact-finding and compiling activity. It provides data upon which management decisions can be made. While good research is expensive, poor research can be even more expensive if it is inaccurate and misleading. No research at all will almost certainly be costly.

For those dealers who think that they need professional assistance, market research, management consulting, and advertising agencies can be beneficial. In most instances, it is unlikely that a small dealer will be able to afford the fees for professional market research. There are alternatives, however.

The Small Business Administration (SBA) sponsors a program called SCORE (Senior Corps of Retired Executives). The SCORE volunteers have expertise in a variety of areas that could be useful to a new dealer. Local SBA offices will provide details on the program.

Many colleges and universities have programs to assist local businesses. Market research could be performed by students under the direction of a business professor. If a nearby university offers a graduate degree program in marketing, most, if not all, of the research could be performed by students who are preparing to enter that profession.

Another alternative is to do it yourself. Extreme care should be exercised when evaluating the data gathered. A rational management decision is needed, not an emotional one.

As a starting place for the do-it-yourself market researcher, the Bureau of the Census, a part of the United States Department of Commerce, has volumes of useful information. Census tables offer per capita income, median family income, and population shifts. If your basic alarm system's price is equal to one-fourth of a typical family's income, it is unlikely that many sales will be made. You will have to design a less expensive system for this market or find another market for your regular system.

Other facts, for example, information about family size, income, and occupation, are available from the Census of Housing for areas as small as census tracts (about 4,000 people) located in 240 different metropolitan regions. Census tracts are comprised of blocks with specific information about the block's population (including minority representation and age ranges), numbers and types of structures, average number of rooms, average value, average rent, and number of units owned and rented. General population data in each tract may be found in the Bureau's publication, *Census Tracts.*

When analyzing census data, remember that they are quantitative and descriptive. Any qualitative judgments or interpretations will have to be made by the person doing the analysis.

MARKET POTENTIAL

Let us consider the two large, general markets: commercial and residential. Granted, there are smaller markets embedded in these large ones; for the time being, however, let's concentrate on the general markets.

Commercial Market

Many alarm dealers consider selling a commercial system to be a difficult task. It is likely that the commercial market will be much more competitive than the residential market.

Educating the commercial prospect on the need for a protective system is not as difficult as educating the residential prospect. Typically, a businessperson already has some idea about his loss prevention needs. What he wants to know from the alarm dealer is whether or not the proposed system will prevent or reduce certain types of losses and if it can be done efficiently, effectively, and economically.

One of the greatest obstacles the alarm dealer or his sales representative must overcome when presenting a proposal to a commercial prospect concerns money; however, the price of the system may not be the problem. More likely than not, the businessperson will evaluate the price of the proposed system and consider the effect the system will have on his profits. Unfortunately, we, as alarm dealers, cannot tell a business owner or manager that by installing our system his losses will be reduced and therefore his profits will be increased. That will probably be the case; however, we cannot prove that it will happen.

Businesses spend money in order to make money (i.e., profit). They are seeking a return on their investments. Again, unfortunately, we are unable to assure them that they will receive a return on their investments in security systems. While a burglar alarm may not prevent a burglary or a fire alarm prevent a fire, either can certainly play an important role in reducing the amount of loss. If the owner knew for sure that he would suffer a loss tomorrow, he probably would be willing to invest in a protective system that would prevent or minimize that loss. Since we cannot predict the future, we cannot make those kinds of assurances.

The alarm dealer who is going to be successful in the commercial market is one who will work closely with the businessperson to design a system that will meet his needs and his budget. Moreover, the successful dealer should be able to explain clearly that while there is no way accurately to measure a return on the dollars invested in loss-prevention equipment and services, certain indicators point toward reduced losses. (Of course, this will mean that accurate, preferably local, data are available for use. Items such as newspaper clippings about fires and burglaries telling about minimal losses and captures at the scene are particularly useful.)

Residential Market

If we consider the concept of "if it is worth stealing, it is worth protecting" as valid, the entire residential alarm market consists of 95% to 97% of the residential structures in the United States. For all practical purposes, it means that almost everyone is a potential residential customer. With a market that large, how can alarm companies fail? Evidently quite easily, considering the number of businesses that close each year.

Oklahoma City, Oklahoma, with a population of just over 400,000, had, at last count, about eighty companies involved in the sale, installation, or service of alarm systems. While the total number has grown during the past few years, almost as many have gone out of business as have come into the industry. For every ten that start up, seven or eight will close their doors.

The obvious flaw in the concept of almost everyone being a potential purchaser of an alarm system is that it is too general. Under the right conditions, almost anything will be stolen regardless of its value. To protect an item from being stolen, however, an alarm system may not be needed. There are other protective methods; locking hardware and patrol or guard services come to mind almost immediately.

Let us assume that we live in an average suburban area with a population of 100,000. In this average city, the average family consists of 2.8 members, which means that there are approximately 36,000 families living in some type of residential structure. If 3% to 5% of these residences already have alarm systems, as national statistics seem to indicate, there are more than 34,000 residences that do not. If there are twenty alarm companies, each capable of installing one system per day, they should be busy for the next six and one-half years. A realistic projection? No. But that is the approach taken by some dealers.

True, the residential alarm market is large and is luring new companies into it daily—and many of them fail. The causes for failure are numerous, though the result is almost always the same: lack of enough profit to sustain operations.

Even with a more realistic view, the potential residential security market is more than large enough to support an alarm industry many times its present size. Sometimes, attempting to take too large of a slice of the pie can be dangerous to the economic health of a business, however. Parts I and II of this book discuss some of the problems that an alarm company may encounter and suggest solutions—or avoidance techniques.

Special Markets

The residential portion of the overall alarm market may have the potential for being the most profitable, but businesses, industries, and institutions must also be considered. Their needs may include specialized electronic systems that monitor temperature or pressure or cycles of machinery. Traditional burglar alarm technology, slightly modified, can expand the market and provide systems to meet special needs.

Farms and ranches occupy a unique niche in the market because they typically combine residential and business security applications and, sometimes, specialized systems. With crime becoming more prevalent in rural areas, this is a market that is as yet untapped.

Similarly, there is more to the market than burglar or, more appropriately, intrusion alarms. Needs exist for fire, hold-up, panic, and medical alert systems. Successful dealers will recognize that they are in the loss prevention industry and not just sellers of alarms.

Fire Alarm Market

Instead of dividing the market into commercial and residential segments, it could just as easily be categorized by the type of alarm system, with burglary (intrusion) and fire alarms being the major groupings.

Fire alarms account for a very small percentage of most professional alarm companies' sales. Why? To be honest, I am not sure. Possibly it is because detailed codes and regulations appear too formidable for the dealer, especially the small dealer. While there are strict specifications for equipment and installation, these should not be discouraging. Alarm dealers who are planning for the future plan a dual approach to protection: intrusion and fire alarms.

Fire alarm systems are not just for businesses; most fire-related deaths occur in the home. The residential market is virtually untapped and is expanding as people realize that there is more to a residential fire alarm than a battery-operated smoke detector.

Already, many insurance companies offer discounts to homeowners who have fire safety equipment in their homes. As losses are reduced even further, increased discounts may be offered, providing a financial incentive for the homeowner to have loss-prevention devices installed.

ENTERING THE MARKET

The decision to enter the alarm market is not an easy one. It should be based on good data. Often, though, good data are not readily available. One of the most common errors of management is the failure to define a problem clearly. A clear solution is difficult when the problem is fuzzy.

A friend who was quite knowledgeable in many areas was rarely at a loss when asked to answer a question. One day, he jokingly remarked, "I can answer any question asked of me. It gets a lot tougher, though, when they want a correct answer."

Correct answers are needed to some very important questions when considering spending the time, effort, and money necessary to establish or maintain a profitable business. Answering the questions listed below will not scientifically prove whether or not you should enter (or stay in) the alarm market, but they will provide some topics that should be considered before a final decision is made. Evaluate and answer each one carefully. If a joint venture is to be undertaken, the questions should be answered by each person involved.

1. Are you familiar enough with alarm equipment to select the items that would provide the protection needed by the customer?
2. Are you familiar with installation methods?
3. Do you have any direct experience with an alarm company or closely related business?
4. Have you ever installed alarm systems or similar products?
5. Do you have a technical background or technical/mechanical aptitude?
6. Have you ever supervised other people?
7. Do you know the competition? Their market area? Products? Specialties? Prices? Reputation?
8. How aggressive are competitors in seeking customers?
9. Are there parts of the overall market that appear untapped? Why?
10. Do you know the *exact* market that you want?

Maybe instead of being the first chapter, this should have been the last. At least, if it were at the end, you would have more information when considering these questions. But here it is, right at the beginning; so, keep the questions in mind as you read the remainder of the book; then come back to Chapter 1, read it again, and answer the questions.

Product Knowledge and Installation Expertise

Familiarity with security equipment is not a prerequisite for entering the market as an alarm dealer; however, success will depend in part on product knowledge. Committing oneself to a program of continuous learning will be invaluable. Recently, alarm technology has been advancing at such a rapid pace that staying abreast of the changes requires a concentrated effort.

Not being familiar with equipment at the outset is not a critical factor if there is a determined effort to learn. Numerous books and magazines are available as resources. Equipment manufacturers and distributors have catalogs, product information bulletins, and technical staffs, and some offer seminars and training programs.

Even a franchised dealer who is unable to select a variety of equipment because of the franchise agreement should be aware of technologic advances. In making a competitive bid, it might be good to know the features, functions, and limitations of your competitor's system.

There is no question that having previous experience as an alarm installer or at least having worked for an alarm company would be beneficial. Again, lack of experience is not critical as long as there is a willingness to learn.

Persons with a technical background, especially those with experience with electronic devices, may find it easier to understand alarm system theory. While it is not necessary to be a technician, an elementary understanding of electronics is helpful.

Even if a new dealer can afford to start his business with an experienced installation crew and technical staff, it is a good idea to have a working knowledge of products and installation methods. Otherwise, how will he know if his employees are working effectively and efficiently?

As a management philosophy, never asking a subordinate to perform a task that has not been performed by the supervisor deserves serious consideration. It is especially important at lower levels of management. Unless the managers and supervisors have actually performed the tasks they have others undertake, they cannot be sure of the true nature of the job. If a supervisor doesn't know how to do a job correctly, how can he know if his subordinates are doing it incorrectly? Understanding installation procedures will be an asset when preparing bids, too.

Management Participation

As a personal philosophy, I believe that it is important for upper management periodically to perform tasks at all levels within the organization. While this is easier for small companies, it should not be overlooked by larger ones. If actual participation is not practical or desirable, top-level management should observe the various tasks periodically.

My own company is small, and as it grows I notice that I have less time to be on the job site. More time is spent in the office with management activities.

Yet, I do not want to lose sight of the two primary functions of my organization: sales and installations. Therefore a portion of my time is set aside for sales calls, either alone or with one of my sales representatives. Another portion is set aside to visit sites and, whenever possible, assist with installations. We must remember that if there are no sales, there will be no installations; and if systems cannot be installed efficiently and profitably, there will be no need to continue making sales.

Knowing the Competition

Keeping track of the competition should be more than just a matter of idle curiosity. We cannot be competitive unless we know our competition, and this means knowing more than simply who the competition is.

What type of equipment are they using? Is it primarily one brand or type? What system features are they selling, that is, what are the customers buying? Are they selling features or price? (Price competition is largely in the mind of the dealer. We take a closer look at pricing and price competition later.)

Are the competitors specializing in some way? Do they seek only residential (or only commercial, institutional, or industrial) customers? Are they only using wired (or wireless) systems? Do they usually install a specific type of space-protection device to the exclusion of others? Do they sell and/or lease systems? Are most of their systems loud local or monitored alarms? In what geographic area(s) are they most active?

What is the reputation of each of the local competitors? What do their customers say about them? What do the local law enforcement and fire officials say? Have complaints been filed with the Better Business Bureau?

If these questions cannot be answered, the competition is not truly known. No one said that obtaining the answers would be easy—just necessary.

Other alarm dealers are not the only competition. Cable television companies, consumer electronics stores, sound and communications dealers, as well as some larger discount and department stores are now offering security products and services. Table 1.1 shows where security directors, managers, and supervisors go to retrofit security equipment and, as it indicates, alarm dealers are just one of several sources.

Burglar alarm equipment can be seen on store shelves and in mail-order catalogs. "One research study recently predicted that by 1983, 42% of the security sales would be in over-the-counter products."[1] As this book is being written, it appears that the prediction is off-target, and actual over-the-counter sales are lagging far behind predicted levels.

Sears, one of the nation's largest retailers, is negotiating contracts with alarm dealers throughout the country. While the terms of the contracts may vary, the

---

[1] Susan A. Whitehurst, "Competing for Your Slice of the Security Market," *Security Distributing & Marketing* magazine, July 1981, p. 33.

Table 1.1 Installation Responsibility

Installation Force	*Total* (%)	*Commercial* (%)	*Industrial* (%)	*Institutional* (%)
Security staff	14.3	15.6	15.5	12.5
Maintenance staff	38.9	24.7	29.6	57.3
Manufacturer	26.6	31.2	33.8	17.7
Alarm dealer	46.3	63.6	42.3	35.4
Electrical contractor	20.1	16.9	21.1	21.9
Other	5.7	2.6	7.0	7.0

Totals may equal more than 100% because system installation may be a joint effort.

Reprinted with permission from *Security World,* February 1981. © 1981, Cahners Publishing Co.

results are the same: local dealers are given the exclusive rights to sell and install alarm systems under the Sears name.

Name recognition and easy financing are definitely in the dealer's favor. Drawbacks to the program include the fact that sales are made by Sears salespersons who have little knowledge of security, and there are limitations to the type of equipment sold and installed.

Still a new venture, Sears' involvement in alarms cannot be judged a success or failure. Only time will tell if significant penetration into the residential market can and will be made.

Market Analysis

When analyzing the market in general, it may appear that gaps exist. There may be certain segments that appear untapped. Moreover, there may or may not be reasons for the gaps.

It is possible that adequate sources of customers will be easily found in certain sections of the market and no one has attempted to enter these areas. Or, maybe they have been entered and found to be unprofitable. Possibly, they are merely pseudomarket segments, that is, they appear to be viable markets but for some reason they are not.

Defining Market Segments. Very careful consideration should be given to selecting the market segment to be entered. Too often, a dealer may not clearly define this factor. It is not enough to say, "I am going to sell alarms." When asked, "To whom?" some dealers respond by saying, "Anyone who needs one." This method of approaching the market is the least desirable.

Unless the dealer can clearly define the market being sought and establish specific goals and objectives, the majority of his efforts may be wasted. Generally, it is advisable, at least in the beginning, to specialize and pursue a specific segment.

There is a story about two men who were discussing the new era of specialization. One of them said, "Did you know that the National Biscuit Company is so specialized that they have a vice president in charge of Fig Newtons?"

An argument ensued and it was determined that the only way to find out for sure was to call Nabisco. The call was made and when the phone was answered, the caller said, "Let me talk to the vice president in charge of Fig Newtons." The voice on the other end came back, "Packaged or loose?"

Become a specialist before becoming a generalist. Learn everything there is to know about selling and installing alarm systems to one market segment before moving on to another segment. If small retail stores are your target, become proficient at those installations before attempting large office complexes. Small mistakes may hurt, but are rarely fatal; larger errors may totally eradicate any hope of profit for several months.

Formula for Success

Not long ago I was told the secret formula for success in business: "Only work half days . . . any twelve hours will do." I am not sure if this is true or not. Nor am I sure that there is any such thing as a formula for success. There are ways, however, to increase the odds of succeeding. Persistence, willingness to learn, and sound management practices will certainly give us a favorable advantage.

THE RESIDENTIAL MARKET—REALISTICALLY

Assume that Anytown, USA, has a population approaching one-half million, with seventy companies involved, to some degree, in the sales and/or installation of alarm systems.

The 500,000 residents of Anytown comprise some 160,000 households of various types (houses, duplexes, condominiums, apartments, and mobile homes). If, as the national statistics indicate, only 3% to 5% of the households have alarm systems, approximately 153,000 of them are unprotected and are potential customers.

At this point, the seventy alarm companies would have about 2,200 potential customers each if the market was divided equally. If an alarm company could reach half of those households, it would take over two years to give sales presentations at two presentations per day.

If 20% of those sales calls resulted in sales, there would be two sales per week requiring an average of two days each to install. If a net (after expenses) profit of $300 per sale were realized, the company's annual net profit would be over $31,000. Moreover, that figure would expand if lease/service revenues were included (and service expenses do not exceed revenues).

Table 1.2 shows the calculations for projecting annual revenues and profits. Accurate, industry-wide figures are not available, but large alarm companies have the capability to predict income with reasonable accuracy. While smaller companies

Table 1.2 Estimate of a Hypothetical Residential Market

	Weekly	*Monthly*	*Yearly*
Sales presentations	10	43.3	520
Sales/installations	2	8.7	104
Revenue ($)	2,800*	12,133	145,600
Net profit ($)	600†	2,600	31,200

* Average sale = $1,400.
† Average net profit per sale = $300.

may not have this same capability, they can nevertheless use some simple estimates, such as those shown in the table, and maintain good records to ascertain whether or not they are meeting their established goals. The key point to consider is establishing goals: if they are realistic and you are willing to work to achieve them, the hypothetical market may become very real.

Remember, too, that the example is only for residential alarm systems; other market segments should not be overlooked.

Elusive Market

With all those potential residential customers, why is there not a booming market? Possibly the biggest reason is that a large-scale national promotion has not been mounted.

The consumption of avocados increased after the National Avocado Association started a nationwide publicity campaign. The same thing happened to milk. Remember the television commercials sponsored by the Dairy Association?

Maybe it is time that the National Burglar and Fire Alarm Association (NBFAA) and the Security Equipment Industry Association (SEIA) joined forces with local and regional alarm associations to educate the public. For such a campaign to work, though, it is going to take the efforts of each of us, working through our local associations. Membership in your local association and the NBFAA is important. Appendix B contains a listing of local, regional, and national associations. Why not take an important step for your future and for the future of our industry and join today?

Getting Started 2

The Small Business Administration defines a small business as one with less than one million dollars in sales or receipts annually. The majority of alarm companies presently in the market are *small* small businesses.

STARTING AN ALARM COMPANY

For a business venture to be successful, the owner/manager should proceed through a series of steps. First (and foremost), planning is essential. This includes establishing goals and the procedures necessary to accomplish them. Setting up a realistic time schedule is important, and recognizing any situation that might affect it is critical. Unless plans and goals are put into writing, checked for completeness, and reviewed periodically, there will be no way to measure progress.

Second, a market analysis (as discussed in Chapter 1) should be performed. As much information as possible should be gathered about the strengths and weaknesses of the competition.

Third, an alarm dealer needs expertise in general business management techniques as well as in the special aspects of operating an alarm company. The business owner should either possess the technical knowledge necessary to operate an alarm company or employ someone who does.

Fourth, when it comes time to add to staff, there is no substitute for hiring competent people. Loyalty and experience in employees are true assets. Moreover, a trusted assistant, or "second in command," may well be the dealer's most valuable asset.

Fifth, do not underestimate the financial needs of the business. In the seeming urgency to get started, many businesses fail to realize what their actual financial needs will be. Businesses that begin on a shoestring frequently find themselves broke.

Sixth, keep records and keep good ones. It is impossible to measure progress accurately without good records. The importance of good record keeping cannot be overemphasized.

Planning

Planning, as a management function, is most important, yet it occupies only a small portion of the dealer's time. Typically, the dealer, especially the small dealer, finds himself enmeshed in daily business activities and neglects to look ahead to next month, much less next year.

A Los Angeles-based research firm, Planning Research Corporation, studied businesses in southern California, where over 1,000 businesses fail each year. Half of those failures are due to a lack of sound and practical planning, according to the researchers.

Most business schools teach planning as part of management curriculum.

The traditional, or scientific, approach is usually taught, and it is worth mentioning here.

Planning should follow a logical sequence:

1. Problem identification: a problem cannot be solved if it cannot be identified and defined.
2. Information gathering: as much pertinent information as possible should be assembled.
3. Information analysis: take a close, analytic look at the information gathered.
4. Alternative solutions: there may be more than one way to solve the problem; make a list of all of them.
5. Solution selection: choose the best solution.
6. Act: implement the plan.

Regardless of the size or complexity of the problem, planning in a logical manner will make it more manageable. Time spent planning will reduce the time necessary to correct more costly problems in the future.

If this chapter looks like something out of a business management text, it could very easily be. The decisions necessary to start a business, any business, are similar. Starting an alarm company is not much different from starting an auto repair shop or a jewelry store.

CAPITAL REQUIREMENTS

Starting an alarm company (a company that sells, leases, and/or installs alarm equipment) without providing a local central station operation would probably require between $15,000 and $25,000 of capital. Purchasing a building instead of renting or leasing one would raise those estimates substantially.

The initial capital investment, which may be a combination of assets including cash and equipment, should be enough to provide for rent, utility and phone deposits, office furniture, equipment and supplies, showroom and/or shop fixtures, insurance prepayments, a service vehicle, and an opening inventory. Do not forget to include the cost of an alarm system to protect the facility. In addition to start-up costs, there should be a cash reserve capable of sustaining business operations for at least a two-month period.

If the amount of required capital necessary to fund a new small business adequately seems large, that is because it is large. A careful and honest analysis of your own capital requirements will reveal your exact needs.

Many who are considering entering the alarm industry will do so undercapitalized: a few will succeed and become successful and profitable; many will not last more than two years.

To those of you who are already established and are successful and profitable after starting on a shoestring—congratulations! To those who are still struggling, there may still be a chance for profitability by applying sound management principles *now.*

Lack of planning for financial needs is one of the most frequent contributing causes to business failures. In addition to knowing where to obtain the money needed, knowing how to use it wisely is important.

Sources of Capital

Small does not equate with simple; that is, do not assume that because a business is small, that it is simple to operate. Many of the managerial problems that arise in large businesses arise at some point in a small one. The most frequent ones will probably be finance-related.

There are several sources of funds for the new alarm dealer:

1. Personal savings
2. Commercial bank loans
3. Trade credit
4. Capital stock
5. Loans from friends and relatives

Investing one's own savings in a business venture combines the risk of ownership with the possibility of profit. Venture (risk) capital is usually the major source of funds, and is supplied by the entrepreneur. Bankers are extremely reluctant to loan venture capital, and if it is available, the interest rates will probably be well above those for less risky purposes.

Commercial bank loans are typically used for the working capital needs of operating businesses. While some initial capital can be found from this source, it may be difficult to locate and probably will be expensive.

Credit that is extended by suppliers is another source of capital. The amount will depend on many factors, the primary ones being current economic conditions and the supplier's confidence in the new dealer. Suppliers tend to be cautious, and rightfully so, and prefer to do business on a prepayment or COD basis until the dealer is established. Let's face it, it is easier to obtain credit once you have a good credit rating; the problem is getting those first few accounts.

While the sale of capital stock is a potential source of capital, it is not a very promising one. The smallness of the enterprise prevents this from being a major source of funds. It is possible, however, that some stock could be sold, and this should not be overlooked as a source. Of course, the sale of capital stock requires that the alarm company be a corporation, which may or may not be the preferred organization structure. (Chapter 3 discusses the forms an organization can take.)

Another source of financing for new dealers is friends and relatives. This source, unless there is none other available, probably should be avoided. Coupling a close personal relationship with a monetary loan may yield friends and relatives who think that they have managerial duties and seek to give advice and/or directly interfere in the business. If this form of financing is used, try to repay the loan(s), with interest, during the first six months of operations.

Additional Capital Needs

As the business gets under way, additional problems of financing and financial management arise. The most frequently stated need of an established alarm company is for working capital. The next is for the purchase of equipment (assets). When considering these needs it is well to keep in mind the fact that most small, new alarm companies operate under a financial handicap; having begun operations with inadequate capital, the same problem will probably exist in the going concern. Owner investment and retained profits are frequently insufficient to expand the business.

The same sources of funds that were available to the new dealer are available to fund an established operation. Personal savings, for example, may have been used to the limit when the business was started. The sale of capital stock probably is still not feasible unless the business is large enough to attract investment bankers. The problems associated with loans from friends and relatives are about the same.

Commercial bank loans and trade credit limits may be expanded after the business has been operating and has established a good track record for repayment. These are good sources for additional capital.

Operating profits retained for the purpose of expansion of the business constitute a major source of capital. Unfortunately, it is rarely adequate for a rapidly growing alarm company.

Business Acquisition

There are three routes to alarm company ownership and operation:

1. Start a new enterprise.
2. Purchase an existing business.
3. Inherit a going concern.

The first method is the one most likely to occur, although there is a possibility that purchasing an existing alarm company is a viable option. Except in rare instances, the last seems unlikely. Our discussion focuses on the first two methods.

There are several reasons why an entrepreneur may decide to start his own alarm company rather than purchase an existing firm. For one thing, it enables the owner to select his own location, products, services, equipment, workers, suppliers, and banker. He can avoid the errors of others, too, such as possible undesirable binding precedents, business policies and practices, and legal commitments of an existing firm.

There are two very good reasons for starting a business similar to other existing alarm companies:

1. There is real, permanent expansion of the market that is not being met adequately by the existing alarm dealers.
2. Existing alarm companies are inefficiently managed, which results in the market not being adequately served.

In some instances, the latter may be the case. For the most part, however, the former reason—market expansion—is predominant.

Pros and cons exist for both starting and purchasing an alarm company. A purchase can sometimes be a bargain, depending on the reasons behind the sale. There are some valid reasons why a business owner would be willing to sell at a price below the business' actual worth. There also is the possibility that the business will be overpriced.

Analyzing the records of an operating company provides better data than estimating the requirements for starting a new company. The new owner may be able to reduce costs or increase sales; either way, profits will be increased.

The seller may, from sentiment or greed, overvalue the business he is selling. Extreme care should be exercised in valuing the prospective purchase. An independent audit or, at least, the advice of your banker, attorney, and accountant should be sought.

An audit will reveal the accuracy and completeness of the financial data. It may show areas where improvements can be made to increase profitability, or it may be that it does not accurately reflect the true financial condition of the business. It would be unwise to accept carte blanche financial statements prepared by the seller's bookkeepers without an independent audit.

Purchasing an operating business has numerous tax considerations. A competent tax accountant should be consulted prior to making a final decision. On the one hand, if the buyer pays cash for the business assets, the purchase price (cost) becomes the tax basis for computing expenses such as depreciation. On the other hand, if he buys the capital stock of an existing company and keeps the corporation operating, his tax basis will be that of the corporation.

Other factors should be considered when taking over an existing business:

1. The extent, intensity, and location of competitors
2. The place the business has in the local market (i.e., its market share)
3. Sufficient number of customers at present and predictions for the future
4. The plans for the community, such as
   a. Zoning changes
   b. Prospective land use for the public
   c. Street or highway changes affecting traffic flow or land use
5. Contingent liabilities or unsettled litigation
6. Mortgages of record against any real property acquired

Before closing the deal, your attorney and accountant should be in agreement. They should have worked closely with you in negotiating the terms of the sale.

If the business is being started instead of purchased, there are still some questions to be answered and some problems to be solved. For most businesses, the three essential factors in success are: location, location, and location. This is not necessarily the case for an alarm company. Few people expect to drive to the local shopping mall and find an alarm store; they do not "go shopping" for an alarm system. (Besides, in order to prepare a bid and recommend a system, the dealer or one of his staff should physically view and inspect the structure to be protected.) So, a high traffic location is not as critical with an alarm dealer as it is with some other types of businesses.

A poor location may prove disadvantageous, though. It would be a poor decision to locate in a section of town where little if any business could be obtained. If the market being sought is commercial, a location near or in the main business district might be a good choice. If upper-income families living in homes valued in excess of $200,000 will be your market, it would be unwise to locate your office and shop in the center of a low-income or subsidized-housing area.

LOCATING AND EQUIPPING AN OFFICE

Alarm installation companies vary somewhat in the services they offer and in their approaches to their customers. The layout of the office and shop area is far less critical for an alarm company than for other types of businesses. The design and set-up should be functional because, except for a few who provide showrooms to display equipment, sales are made and services are performed on the customers' premises.

When considering general office space, there should be a minimum of sixty-five square feet per worker. Desks and files should be arranged so that they are convenient and easily accessible. There are other considerations, too. How much total space will you need? Will a small office be adequate? What kind of storage space will be required? How fancy must the office be? Who will be seeing it? Will clients come to the office? If so, for only a few minutes or for lengthy meetings? What, if any, special features will be needed? Sink? Numerous electrical outlets? Special ventilation?

Once you have decided on the basic furniture, the next major purchases you have to consider are office machines, like a typewriter, adding machine, calculator, and maybe a copying machine and computer. Although tax laws are constantly changing, there seems to be a trend toward allowing businesses to deduct a certain amount of the expenditures for office equipment without using depreciation formulas.

Office Supplies

Office supplies are not the first things to come to mind when planning a business. They should be planned for, though, and they can be rather expensive. Consider that, as a new business, you will need a variety of items: pens, pencils, erasers, typing paper, correction fluid, scratch pads, paper clips, rubber bands, stapler and staples, scissors, tape, stamps, several sizes of envelopes, mailing labels, clip-board, files, filing supplies, typewriter ribbons, and so on.

In addition to these basics, some thought should be given to creating your company image on paper through letterhead stationery and business cards. White, buff, gray, manila, and tan are the most acceptable colors for business paper. If in doubt, choose white.

While a logo and fancy letterhead have a certain appeal, do not overdo it. The idea is to make someone remember you (and your company) by remembering the logo. Elaborate type and overdesigned logos may make the letterhead difficult to read and defeat your purpose.

BUSINESS SIZE AND RELATED PROBLEMS

There are a few problems faced by small alarm companies that are caused by their size and the unique aspects of the industry. One very important problem is the shortage of time that the entrepreneur can devote to true management tasks.

The functions of management are to plan, organize, direct, coordinate, and control. In a small company, these responsibilities usually belong to a single person: the owner. It is likely that he also helps out with installations, or, depending on the size of the company, does all of the work himself. Even though the owner does not function solely as an executive or decision maker, management functions must still be performed if the business is to be successful.

Another problem that frequently interacts with the time pressure problem is lack of specialized management skills. The owner may be an excellent technician or installer; he may possess outstanding knowledge of security products and techniques. Unless he is able to perform the management functions or can rely on an associate for managerial assistance, however, serious doubts about success are raised. If desire for success could be transformed into technical or managerial ability, there would be fewer business failures. Unfortunately, such transformations cannot be accomplished.

Small alarm companies experience difficulty in hiring and retaining staff members at all levels within the organization. Larger companies' recruiting activities and remuneration plans usually attract qualified applicants. Many who start with a small firm will leave after gaining some experience and seek employment with larger businesses that offer advancement opportunities.

If it seems that a gloomy picture has been painted for the newcomer considering joining the ranks of the alarm industry, it is because certain aspects of the business should be recognized and understood. Awareness of the potential for success in a rapidly expanding market should be weighed against the possibility of failure. While there is no substitute for experience, some experiences can be costly, even disastrous. A tremendous market exists and will reward those who enter it in a knowledgeable and professional manner.

Organizing the Business 3

Once you have made the decision to enter the alarm industry, the next step is to decide on the legal structure that will be most advantageous to you. Even if you already own a company that has been operating for some time, it is possible that you might want to consider changing its legal structure.

LEGAL STRUCTURES

The kinds of legal structures are:

1. Sole proprietorship: a business owned and operated by one person
2. Partnership: two or more persons as co-owners
3. Corporation: a legal entity separate and distinct from its owners

Each has advantages and disadvantages that should be considered carefully before a decision is made.

In a sole proprietorship, the principal owns all of the assets. The owner/proprietor has the legal right and exclusive title to everything. He also is responsible for any liabilities incurred by the firm.

Uniting with others to form a partnership means that profits and losses will be shared. Several types of partnerships exist, with general and limited partnerships being the most common.

A corporation is like an artificial person. It exists because of the laws that allowed it to be created.

Sole Proprietorship

The advantages of sole proprietorship are its ease of formation and the fact that the owner is entitled to all of the profits as well as all of the decision-making processes. It is the most flexible type of legal structure and also the one that is most free of government control and taxation.

The greatest single disadvantage is unlimited personal liability, which extends beyond the business operation. You, personally, are liable for debts should the business be unable to meet its obligations. In other words, if claims are made against the business that it cannot pay, your savings, house, or other personal property may be used to satisfy them.

Less money may be available in the form of bank loans and lines of credit, especially for alarm dealers whose operation is primarily service with little inventory to use as collateral. Long-term financing may be extremely difficult to obtain.

Partnership

A general partnership is a voluntary association of two or more people. It is agreed that each party will contribute money, efforts, and/or skills and will share the profits or losses proportionately. Similar to the sole proprietorship, unlimited liability is created for partnership obligations.

In a limited partnership, liability of some partners may be limited if they do not actively engage in the management and operation of the business. Only the general partners in the limited partnership have unlimited liability.

As a business form, partnerships offer much the same flexibility as a sole proprietorship. Two or more persons acting together are often stronger than if they acted alone. Shared management responsibilities and more experience and skill from which to draw are some of the advantages.

Partnerships are formed relatively easily. It is recommended, however, that an attorney be consulted to explain the various types and prepare the contract. The Small Business Administration's (SBA) pamphlet, "Selecting the Legal Structure for Your Firm," also offers worthwhile comments on the forms of partnerships. A copy can be obtained by contacting the nearest SBA office.

Corporation

In 1819, Supreme Court Justice Marshall defined a corporation as "an artificial being, invisible, intangible, and existing only in contemplation of law." Simply stated, it is a method of business organization that is distinct from the person or persons who own it. Incorporating a business is not always easy. Moreover, some variations in corporate structure exist.

Close corporations allow shareholders greater flexibility in the management of the organization. It allows the relaxation of certain rules and regulations.

Subchapter S corporations, as defined by the Internal Revenue Code, allow shareholders to be taxed as partners in a partnership. While the shareholders are taxed instead of the corporation, there can be a significant drawback: shareholders may be taxed on income they do not receive because of the inability of the corporation to distribute cash.

Some double taxation results from a corporation's dividends. The corporation, as a separate entity, files its own tax return and pays taxes on profits. The dividends distributed to shareholders are taxed as part of the corporation's profits and are taxed again as income for the shareholder.

The advantages of a corporation are limitation of liability, simplicity in transferring ownership, stability, and ease in securing financial backing. The primary advantage probably is the limitation of liability.

Some of these advantages are relative. A corporation does not ensure stability and, if it is not profitable, transfer of ownership may not be simple. Financial backing requires a history of profitability and good management; in the case of a new business, potential backers must be convinced that it will have good management and will be profitable.

It should be noted, too, that the advantage of liability limitation may be eroded. In some cases, especially for small or family-owned corporations, the shareholders may personally have to guarantee corporate obligations.

Table 3.1 Advantages and Disadvantages of Different Organizational Structures

Structure	*Advantages*	*Disadvantages*
Sole proprietorship	Low start-up costs Freedom from regulation Owner in direct control Minimal working capital Tax advantages to small operations All profits go to owner	Success requires owner's constant attention to business Difficult to raise capital Unlimited liability
Partnership	Ease of formation Relatively low start-up costs Partners provide additional capital sources Varied management and technical skills Possible tax advantage to partners Limited regulation	Unlimited liability Divided authority Shared profits Additional capital sources may be limited Must locate suitable partner
Corporation	Limited liability Potential for specialized management Ownership transferable Sometimes easier to raise capital Status as legal entity Potential tax advantages	Closely regulated Expensive to start Double taxation on some income Detailed, extensive record keeping May be restricted by charter

The cost of starting a corporation is higher than that of a sole proprietorship or partnership. In addition to the attorney's fees, special taxes and charges for permits or licenses may be required. The cost of incorporating, not including the initial capital investment for stock, probably will be $750 to $1,500.

Table 3.1 shows a summary of the advantages and disadvantages for the different organizational structures.

PROFESSIONAL ADVISORS

While the success of your business will depend on a number of variables, there are certain steps that can be taken to improve your chances. One is the selection of professional advisors. In particular, consider your organization's need for an attorney, accountant, and insurance agent.

Some small businesspeople consider the use of professional advisors as a sign of extravagance. Actually, it depends on need. Competent advice can prevent costly mistakes. Even though attorneys and accountants are not inexpensive and insurance premiums always seem to be rising, investment in these services may prevent costly litigation, tax problems, or liability claims.

You and Your Attorney

If you have not worked closely with an attorney in the past, you may be in for a few surprises. Although the small-businessperson and the typical lawyer consider themselves in different professional leagues, they have much in common. Both operate in a market where they are free to choose their own customers and clients, and both are faced with the problems of running a business.

There is nothing wrong with asking about an attorney's credentials and fees. Your customers will be asking the same questions of you as an alarm dealer. Ideally, your attorney should be knowledgeable about the legal issues of operating an alarm company and should be up to date on current legislation affecting the industry. If you are not sure whether an attorney is versed in the pertinent aspects of the law, ask your local alarm association for the names of others in your area who work with alarm companies.

Specific points to consider when selecting an attorney are discussed in detail in Chapter 19. For now, be aware that sooner or later you probably will need legal advice. If you have a working relationship with an attorney who understands your company's operation, you may have a better chance of avoiding or winning a law suit.

The Need for an Accountant

The quality of your financial advisor's advice directly affects your company's taxes, balance sheets, and possibly, management decisions. Whether formally or informally, someone provides financial advice to you. It could be your partner or your bookkeeper, or you could be responsible for the accounting tasks. Remember, though—the quality of that advice is very important.

A professional public accountant could be your company's most important advisor. A long-term working relationship with such a person means that he will do more than review the books and compute taxes; he can provide crucial advice on cash flow, credit, and other financial matters.

The Company Bookkeeper. Most small companies have limited accounting staffs, with the majority employing only one person: a bookkeeper. In many small dealerships, the owner or manager is the one who acts as bookkeeper.

As your company grows, more people may be added to the accounting staff. Even then, unless your business is very large, a public accountant is probably your best financial assistant.

Information Source. Preparing tax returns and financial statements may be your accountant's primary responsibility, but he should provide other services as well. It is his job to stay abreast of tax law changes and to keep you informed. The reports and tax returns that he prepares should satisfy the Internal Revenue Service and they should be made into useful financial tools for management.

Instead of considering your accountant as a part-time manager of the accounting "department," consider him as a full-time advisor to management. You should feel free to talk with your accountant whenever the need arises. That may be only once a month when he picks up your books or it could be daily during a cash flow crisis. Moreover, communication is not a one-way street, going from you to your accountant. If he believes that he has information that will benefit you, he should initiate the call. After all, what you are paying for is his professional advice, isn't it?

Selecting an Accountant. If your annual sales are $25 million or more, you might want to consider one of the "big eight" multinational accounting firms. If your sales are not quite to that level yet, a local accountant will probably be able to provide all of the services that you require.

The key to selecting an accountant is to locate one who knows the alarm industry or at least is familiar with similar service-oriented businesses. If you will be leasing alarm systems, it would be helpful if he has a background that includes that aspect of your operation.

Professional Fees. Fees for services depend on the size of the accounting firm and the services required by the alarm dealer. Typically, it could be expected that it would cost at least $100 per month for an accountant who provides only very basic services. A more realistic fee would be about $250 per month for one who works closely with his customers as an advisor as well as a preparer of tax forms and financial statements. A good accountant will probably earn his own fee by saving you money that might otherwise have been spent on taxes. Viewed from that perspective, a good accountant costs nothing.

Your Insurance Agent

Many of the same criteria for selecting an attorney and an accountant should be used to select an insurance agent. In addition to the usual types of insurance that can be provided by local agencies and brokers, there are some specialized types especially for alarm companies.

Types and Coverages. General liability insurance is the first type to be considered. It is frequently referred to as BI and PD—bodily injury and property damage—by those in the insurance industry. General liability covers you against liability for normal business operations.

Included should be a provision for "care, custody, and control" coverage, which protects you in case you bring an item back to the shop and break or lose it while it is in your care, custody, and control. Many policies specifically exclude this type of coverage; check yours to be sure it is included.

Another important area to be covered is employee dishonesty. Should one of your employees be proved to have stolen something from a customer, you need coverage for that third-party liability.

Part of your insurance package also should include owner, landlord, and tenant liability coverage—OL and T. If an accident occurs on your premises, you could be held liable, but OL and T coverage protects you against that type of claim.

All of these coverages are needed regardless of the type of business. Alarm dealers, however, should have additional coverage.

The first to consider is manufacturer's and contractor's liability coverage. When one of your installers breaks a window or drills through a water pipe, you can count on a claim being filed.

Completed operations coverage assures that you will be protected if a system is installed and, for whatever reason, does not get totally connected. There are numerous causes for completed operations claims.

Products coverage includes damage or injury due to product failure. First, do not assume that the manufacturer has product liability insurance; he may not. A rather unusual liability is created because the alarm dealer may take the products of several manufacturers and combine them into one system. Second, product liability policies do not cover the failure of systems, just the failure of specific products.

Most insurance policies specifically exclude errors and omissions coverage, so alarm dealers should have a specific policy written. For example, if you install a system and overlook wiring in an accessible window, through which a burglar enters, you could be considered to have omitted important coverage and could be held liable for any loss.

Since we as alarm dealers are specializing in protecting people and property, we must protect ourselves as well. Part of that protection consists of a good insurance package.

LIMITING LIABILITY

The number of lawsuits brought against alarm companies has risen dramatically over the past few years, increasing 600% during 1980 and 1981. Limiting liability is a joint effort between the dealer and his staff, his attorney, accountant, and insurance agent.

Self-Protection

If the alarm systems that you install are 100% effective 100% of the time, questions of liability should not arise. If, however, a component or system fails occasionally and a loss occurs, you could be held liable.

One of the first questions asked when discussing liability is, "What if the customer refuses to accept a system that will protect all possible points of entry? Am I still liable?"

Unfortunately, you could be. To avert this possibility, you should have a good contract that is used with every customer. Your attorney should play a major role in developing such a document and your insurance agent and insurance carrier should be aware of the contract's provisions and offer their suggestions. Moreover, any time a customer contracts for an alarm system that is less than complete, he should be advised, in writing, of the deficiencies and of your suggestions for making the system more secure.

Gentlemen's agreements and handshakes may be acceptable for some types of businesses. They definitely are not acceptable for those of us who are in business to protect lives and property. There is no substitute for a good contract to help limit liability.

FINANCES

In Chapter 2 we discussed some methods of financing a business. Included here are a few suggestions as to how to locate funds and how to ask for the money.

If you think about it, the primary product that bankers have to sell is money. Why, then, are there so many obstacles to obtaining a loan? It should be remembered that bankers are a typically conservative group, and even though they are willing to loan money, they have some very stringent federal and state regulations to abide by. To overcome these obstacles, you should start with a well-prepared loan proposal. Some specific tips for preparing a proposal are discussed at the end of this section.

Basically, there are twelve types of bank loans; each is described briefly. Not every one may apply to your particular operation.

Short-Term Loans

Technically, a short-term loan is for less than a year. In some cases, it may overlap a medium-term loan and last for two or three years. Small alarm companies frequently use short-term loans to finance receivables or inventory.

Line of Credit. The line of credit loan consists of a specific sum that is set aside at the bank and upon which the borrower may draw as needed. While interest is only computed on the amount actually drawn, a commitment fee of ½% to 1% of the total credit line usually is imposed. Some banks may waive the fee in favor of a compensating balance, a sum that must be kept on deposit throughout the loan period.

A nonbinding line of credit may be negotiated, though it will contain no guarantee that money will always be available. Depending on the economy and your business' financial position, it may be withdrawn at any time.

A committed line of credit assures you that the money will be available, but a higher commitment fee probably will be imposed. Inserting the word "committed" in the loan agreement assures the availability of funds.

Short-term lines of credit must be cleared periodically. That is, under most banks' rules, you must be fully paid up for thirty days a year. If this presents a problem, some banks offer a revolving line of credit. It involves an annual review and renewal, but does not require clearing your account.

Inventory Loan. The usual inventory loan runs six to twelve months and requires the same thirty-day clearing period as a line of credit. From the dealer's point of view, an inventory loan is very similar to a line of credit.

Instead of issuing a formal line of credit, especially for larger amounts of money, some banks prefer to write a short-term loan to carry inventory. The collateral may be the inventory itself.

Commercial Loan. Commercial loans require no installment payments; they are simply paid back in a lump sum at the end of the term, typically three to six months. Bookkeeping is reduced for the lender and the borrower.

While commercial loans may be used for any purpose, many alarm dealers use them to finance inventory. Careful scrutiny of the borrower's credit rating is made before a commercial loan is approved. The bank's main concern is how the borrower will be able to pay back the entire lump sum.

Accounts Receivable Financing. Many alarm companies find that receivables tie up large amounts of working capital. Some banks, however, will not agree to accounts receivable financing for small amounts. It would be unlikely to find such financing for amounts of less than $100,000 per year.

Due to the unusual nature of an alarm dealer's business, with his lease and service agreements, approaching a banker with such a proposal might be profitable. A well-prepared loan proposal and a history of success in the alarm business might cause a bank to offer the financing for lower amounts.

Factoring. Factoring is similar to accounts receivable financing, except that the bank buys the receivables outright. To use this method, the dealer would have to sell and finance alarm systems instead of leasing them.

Medium-Term Loans

Medium-term loans, usually from one to five years, are more likely to require collateral than short-term loans. Medium-term loans are frequently used for shop equipment, furniture, fixtures, and expansion. Although they can be used to purchase assets, the bank may or may not view new assets as collateral.

Term Loan. Most term loans provide for 80% to 90% of the cost of the asset to be purchased. Many are written for five years or for the useful life of the asset. Typically, repayment is made in quarterly installments of principal plus interest. Principal payments remain constant, with interest payments declining over the term of the loan.

Monthly Payment Business Loan. If the large quarterly payments of a term loan appear difficult to repay, the monthly payment business loan offers a modified payment plan, permitting you to make approximately equal monthly payments over the entire period.

Long-Term Loan

Long-term loans of five or more years are not used as frequently as short- and medium-term loans, and they are more difficult to obtain. They are commonly used for real estate purchases, major expansion, acquisitions, and starting capital for a new business.

Commercial and Industrial Mortgages. Commercial and industrial mortgages may be written in a variety of ways, depending on the value of the building to be purchased, your company's long-range profit projections, and the bank's policies. If you are lucky, you may get a twenty-five-year mortgage; more likely it will be five to ten years.

Monthly installments for a five- or ten-year loan may be geared to a fifteen- or twenty-year mortgage, but at the end of the loan period, you will be faced with a large, lump sum, "balloon" payment. You may be able to refinance the entire amount still owed, but there is no guarantee of this.

Real Estate Loan. If you already own real estate and want to borrow against its value to finance expanding your business, you may be able to add a second mortgage. Sufficient equity and a good credit standing are necessary.

Personal Loan. Many bankers believe that an owner's personal assets should provide much of the financing for major expansion projects, so you may have to think about including a secured personal loan in your long-term financing plans. Any property that you own can be used as collateral—marketable securities, savings accounts, and certificates of deposit.

Start-up Loan. Bankers consider start-up loans to be similar to expansion projects, and many of them believe that much of your own money should be used. In addition to funds raised by personal loans and partners' investments, you may be able to obtain a term loan from the bank's venture capital funds through a Small Business Administration guarantee. Unless your bank participates in the SBA's Certification Program, it can be a rather involved ordeal.

SUGGESTIONS FOR PREPARING A LOAN PROPOSAL

A loan proposal usually consists of eight parts. Most of the information required already should be on hand. Just remember that you are trying to impress a banker.

Summary

On the first page of the proposal, give a summary of what the rest of the document contains. Include your name and title, your company's name and address, and the nature of your business. Also include the amount of money that you are requesting and explain how it will be used. Finally, and very important, tell how it will be repaid, for example, increased sales, more service contracts, more leased systems, or whatever activities will make your company capable of repaying the loan.

Business Description

Describe your company's legal structure. Tell how long you have been in business, the number of employees you have, and the union status of the company. List your current business assets.

Describe your products and services. Define your market. Identify major customers and competitors. Describe your inventory and indicate its average size and turnover. Report on your accounts receivables and payables. (Receivables should be current; too many delinquent accounts may cause a negative reaction.)

Top Management Profiles

In two or three paragraphs, describe the education, background, experience, and skills of each of your key people. (Experienced management is a positive factor.)

Projections

Using your current market share as a base, explain your opportunities for growth for the next year and the next five years. List a timetable that shows the various stages of growth and the specific goals that will be attained at each level.

Financial Statements

In addition to a current balance sheet and income statement, include your financial statements for the past three years. Project balance sheets and income statements for the next three years. The financial statements should be audited if possible. If you cannot afford a full audit, ask your accountant to review and comment on the financial statements. (Prepare two sets of financial statements for the projections: one with the loan and one without.) Including your own personal financial records, with tax returns for the past three years, is suggested.

Purpose

Be specific with your statement of purpose. Why do you want the loan? Vague, general statements should be avoided. Explain exactly what you intend to do with the funds.

Amount

Ask for the exact amount needed to achieve your purpose. Whenever possible, include cost estimates for products and/or services the loan funds will purchase. Support the amount you are requesting in as many ways as possible.

Repayment Plans

The repayment plan may well be the most important aspect of your proposal. The banker wants to know if he will get his money back. Keep in mind several banking criteria for loaning money:

1. The asset you want to finance must last as long as the loan period.
2. The asset should generate the repayment funds (i.e., by increasing sales, reducing costs, etc.)
3. Your projected balance sheet should clarify your company's ability to meet interest as an expense and repay principal from net profits.
4. Show that you have at least two ways to repay the loan. Should the first way (the asset) fail to generate loan repayment capital, another way should be available.

Fortunately, the bank's lending officer will assist you with part of the proposal; however, all of the financial data will have to be provided by you and your accountant.

The bank is in business, too, so do not limit yourself to the loan department. There are other banking services that you can use. Refer other people to the bank as depositors. If you are interested in the bank, maybe it will be interested in you.

TAXES

Another important topic to consider when planning and organizing a business is taxes. Granted, it is not a pleasant task, but if we are in business and making a profit, we will have to pay taxes.

Estimating Taxes

High interest rates and new tax laws make the drudgery of quarterly estimated tax payments a bit more exciting. Pay too much and you may have to borrow money at a high interest rate from the bank when you could have kept it for free; pay too little, and you will be subjected to interest and penalties from the Internal Revenue Service.

Companies are supposed to pay one-fourth of their total tax bill in each of four estimated tax installments. If you fall behind in estimated tax payments, however, you cannot catch up and escape a penalty. The IRS is very efficient when it comes to assessing penalties. If your company is a corporation that has more than $1 million in taxable income, the tax rules get even tougher.

Managing the Organization 4

What is management? What functions do managers perform? While these are involved questions that are explained in more detail in other parts of this book, we still must have a concept of the answers before much of the material will be understood. We start, therefore, with an exploration of what management is; then, as this chapter progresses, we explore the functions of management. Concluding the chapter is a discussion of an important planning technique for managers.

MANAGERIAL ACTS

If we were to follow a manager about all day and write down everything that he does, the list would probably look somewhat as follows:

Talks to employees
Gives directions to lower-level supervisors
Dictates letters
Establishes installation goals
Hires new installer
Reads mail and reports
Attends meeting
Makes decision about new alarm equipment
Plans for new market entry

These activities are both physical and mental. The physical activities revolve around the concept of communications. The manager is either telling someone something verbally or in writing, or he is receiving a communication by the written or spoken word.

His mental activities, however, cannot be observed directly, but we know through his communications that he thinks and makes decisions. The ultimate objective of both physical and mental activities is to create an environment in which other individuals willingly participate in order to achieve objectives.

MANAGERIAL FUNCTIONS

Management functions are categorized differently by different managers. Basically, there are five general, separately identifiable functions.

Planning

To some extent, every manager must make plans. They vary from immediate tasks to long-range objectives, from simple to involved, and from departmental to company-wide impact. For example, a manager may plan the work for tomorrow, decide when vacations will be taken, decide which installations will be started next week, or determine the company's growth pattern during the next five years. Planning is nothing more than looking ahead, a vital function performed by every manager. Determining future activities necessarily involves a mental look ahead with recognition of needed future actions, whether they be performed tomorrow or next year. It involves conceptualizing about future events and making decisions today that will affect the future.

If future events could be determined with accuracy, a plan of action could be developed to accomplish the objectives of the firm under the conditions the future would bring. The future, however, is not certain, and at best, forecasting is a game of educated guessing.

Therefore, the manager estimates or forecasts that one of several possible conditions will exist at some given projected time. He actually develops a series of plans, some of which will be put into effect depending on the conditions existing at that time.

Planning is not a function reserved exclusively for top management. On the contrary, it is one of the functions that every manager performs, regardless of his location in an organization. The higher the level, the more time spent planning.

Organizing

A manager must organize—people, materials, time, and jobs. By so doing, he creates an environment that will be conducive to achieving the organization's objectives. Organizing consists of:

1. Determining what activities need to be done
2. Grouping and assigning these activities to subordinates
3. Delegating the necessary authority to the subordinates to carry out the activities in a coordinated manner

When you, as a manager, direct work, establish goals, and affix authority relationships, you are performing organizing functions in addition to planning functions. Before you can organize, you must plan; neither function is clearly or separately distinguishable, but both are intermixed in the overall management function.

Directing

In addition to planning and organizing, a manager must succeed in directing the activities of others. This function deals directly with influencing, guiding, or supervising subordinates in their jobs. Directing cannot be performed alone; it must be executed with planning and organizing. An unplanned or disorganized directive is useless.

Coordinating

Few tasks can be undertaken without coordinating the efforts of several people, inside and outside the organization. A typical alarm installation may require coordinating the efforts of the installation crew, the equipment supplier, the city inspector, and the telephone company.

It is the manager's job to be sure that the various tasks are scheduled and implemented in an efficient and economical manner. As with the other functions, it is difficult to isolate coordinating. It, too, is part of the composite that includes planning, organizing, and directing.

Controlling

Whenever people are joined together in a common undertaking, some form of control is necessary. Orders may be misunderstood, rules may be violated, or objectives may unknowingly shift. Whatever the reason, it seems that the larger the number of individuals concerned, the greater are the probabilities that inappropriate action (or no action) will be taken.

Controlling merely consists of forcing the undertaken tasks to conform to prearranged plans. Thus planning is necessary for control. Similar to the other functions of management, controlling is not performed separately. That it cannot exist alone is immediately apparent when we realize that managerial control consists of preconceived and planned acts that must have organization. Similarly, employees must be directed and activities coordinated for control to exist.

Summary of Management Functions. The management process is not a series of separate functions (planning, organizing, directing, coordinating, and controlling) that can be performed independently. It is a composite process made up of these individual ingredients.

If you have any doubts about the composite aspect of the functions, consider this:

> A *plan* is a course of action, an *organized* scheme for doing something. Planning without organization, therefore, is impossible. A plan must be known and communicated. To effect this communication, some *directions* must be given and some *coordination* must occur. To *control* is to verify something by comparing it with a standard (the plan) and taking action if necessary.

THE SUCCESSFUL MANAGER

The job of a manager is to create an environment conducive to the performance of acts by others so as to accomplish personal as well as company goals. Managers should be able to inspire, motivate, and direct the work of others.

Characteristics

A truly definitive list of characteristics for a successful manager would be impossible to develop. The list below includes some of the most desirable ones:

1. A manager should be able to think. Specifically, he should be able to think clearly and purposefully about a problem.
2. A manager should know how to express himself clearly. His chief physical act is communicating. The best-conceived idea is worthless if it cannot be communicated.
3. A manager must possess technical competence. He does not necessarily have to be a technician, but he should have the technical ability necessary to enable him to manage effectively.
4. A manager should be able to think broadly. Broad comprehension is necessary to see the effect of each proposed action on the whole organization.
5. A manager should be a salesman. Selling an idea, convincing others of its worth, is one of his major tasks. Selling a plan of action is a vital part of communication and motivation.
6. A manager should possess moral integrity. Both his superiors and his subordinates should have implicit confidence in him and his actions.
7. A manager should be emotionally stable. He should keep his personal feelings out of business problems.
8. A manager should have skill in human relations and insight into human motivation and behavior. This enables him to lead, not drive, his subordinates.
9. A manager should possess organizational ability. A logical, ordered process is invaluable in achieving established goals.
10. A manager should be dynamic. This is a characteristic trait of leaders.

Leadership

Leadership is a quality that inspires others to perform. It is an aspect of a manager's personality that enables him to influence others to accept his direction freely or willingly.

A good leader is not necessarily a good manager, but an effective manager must have many of the qualities of a good leader. A leader must be able to recognize each of his follower's needs in order to motivate him through his needs. Following are some general qualities that good leaders exhibit:

1. The desire to excel: a leader is never content with second best; he must always be in the lead. He must be a self-starting individual who is willing to engage in long hours and hard work to achieve success.

2. A sense of responsibility: a leader is not afraid to seek, accept, and faithfully discharge responsibility.

3. A capacity for work: a good leader is willing to shoulder the demands of success—long hours and hard work.

4. A feel for good human relations: a leader studies and analyzes his followers, trying always to understand them and their problems. The ability to understand his fellow workers is probably the most important single characteristic of a good leader.

5. Contagious enthusiasm: a good leader should impart a positive, forward-looking attitude to his associates.

Obviously, these are not the only qualities of good leadership. Intelligence, character, integrity, and other such traits are equally important.

TIME MANAGEMENT

Everything we have discussed so far about management involves time. Time management may well be the most difficult of all of these tasks.

Time management consultants often refer to the 80/20 rule, which reflects their discovery that people tend to spend about 80% of their time on tasks that produce 20% of the results. Too many people work diligently on low-value activities. These activities may make you feel good and give you a sense of accomplishment, but they are not significantly productive.

Time Management Tips

Here are some general hints on managing your time.

1. Figure out at what time of the day you are most productive and make sure that you schedule important work during those hours. If, for example, you are a morning person, schedule less demanding tasks for the afternoon.

2. Keep a detailed log of how you spend your time. You will quickly see when and how you waste time, and you will probably be able to spot your most productive hours if you are not aware of them.

3. If the log shows lack of self-discipline, create new time-management habits. Write a fairly rigid schedule for yourself and stick to it until it becomes habit.

4. Keep a calendar, preferably covering a week at a time, so you can always see what you have to do.

5. Do similar tasks at one time; for example, make all of your telephone calls or write all of your letters at one time.

6. Relegate the small or routine tasks to your least productive hours. (This is the time to write letters and make phone calls.)

7. Get someone else to do work you do not absolutely have to do. If your secretary or assistant can perform the task, delegate it.

8. Use downtime—when you are riding a bus or waiting for an appointment—to do certain routine or easy tasks, such as reviewing a memo, figuring your expense account, reading the morning mail, or reading trade journals.

9. Resist the urge to handle the mail as soon as it arrives. Save it for the less productive time that you have scheduled.

10. Control paper. Keep your records simple and look for ways to streamline.

11. Keep things where they belong and keep them in logical places.

12. Eliminate unnecessary meetings.

13. Establish a time for planning (quiet time) and handle only true emergencies, should they arise, during that period.

14. Try to make your first hour at work your most productive hour.

15. Attach priorities to tasks. Do not spend more time on a project than it is worth.

16. Jot down notes of things that need to be done. Do not try to do them immediately.

17. Use your note pad for making notes. Do not attempt to rely on your memory for important information.

18. Keep unscheduled and social visits to a minimum.

19. When someone brings you a problem, expect him to have a suggested solution in mind.

20. Do a job right the first time so you do not have to do it again.

Time management gives managers more time to manage and perform the five functions presented earlier.

PERT

The acronym formed from the first letters of *P*rogram *E*valuation and *R*eview *T*echnique is PERT; it is an approach to planning, organizing, directing, coordinating, and controlling the work efforts required to accomplish an established goal. It can be applied to projects of any size in any field of endeavor and is particularly useful in analyzing the different activities that are part of an alarm system installation. While PERT will work with any size of installation, the larger the job, the more valuable it is.

The definition of PERT is "a statistical technique—diagnostic and prognostic—for quantifying knowledge about the uncertainties faced in completing intellectual and physical activities essential for timely achievement of program deadlines. It is a technique for focusing management attention on danger signals that require remedial decisions, and on areas of effort for which trade-offs in time, resources, or technical performance might improve capability to meet major deadlines."[1] The key words in the definition are *focusing management attention.*

In PERT, management has a sound solution to a long-standing problem: planning and controlling complex activities. The installation of a security or fire alarm system is an ideal application for PERT, which was originally developed by the Special Projects Office of the Navy for the Polaris program.

PERT Technique

The main objective of PERT is to furnish information for managerial purposes. It is not a decision-making process; it simply generates useful information for planning and control. At the beginning of a new project, an alarm system installation, PERT focuses attention on:

1. The tasks that must be done in order to complete the job
2. The sequence, timing, and in some cases, costs of performing the tasks

When the installation reaches the development stage, PERT begins to generate control information concerning:

1. The status of task performance
2. Available alternatives to compensate for missed schedules

To use this technique effectively, an alarm company manager must have a basic understanding of the principles on which it operates and the appropriate method of application. He also must know the nature of the data that PERT furnishes and how they can be used in making project decisions. For the remainder of the chapter we attempt to provide this type of understanding.

PERT Network

The first step in the PERT process is to define the particular goal or target that the program under consideration is designed to achieve. For our purposes, we will assume that our goal is the installation of an alarm system. Then, a network is constructed; this is simply a schematic model depicting a sequential work plan to accomplish the goal. This network, or model, traces the major performance milestones, or events as they are called, and the activities that connect them. The basic relationship between events and activities is shown in Figure 4.1.

An event is a significant specific accomplishment (physical or intellectual) in the program plan, recognizable at a particular instant. Events do not consume

---

[1] Willard Fazar, "Progress Reporting in the Special Projects Office," *Navy Management Review,* April 1959, p. 2.

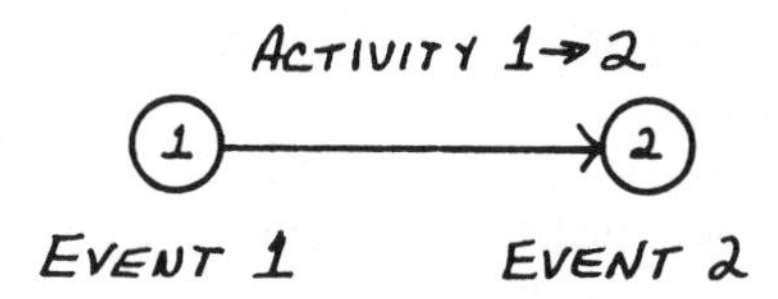

Figure 4.1 Diagrammatic relationship between events and activities in a PERT system.

time or resources (manpower, capital, equipment, etc.) and are represented in the network by circles. An activity represents the work or expenditure of resources necessary to move the project from one event to the next. In Figure 4.1, for example, the activity 1 → 2 might represent the total time or work required to install a perimeter protective circuit.

When designing a PERT network, the manager should keep in mind the following cardinal rules:

1. Events must take place in a logical order.
2. The work and time it takes to get from one event to another are denoted by an activity (→).
3. No event can be reached until all the activities leading to it are completed.
4. No activity can be completed until the event preceding it has occurred.

To demonstrate network construction, let us see how we might set up a simple PERT chart for the work required to install an alarm system. A list of the major milestones in this process can be drawn up as follows:

1. Contract/proposal completed
2. Preinstallation planning started
3. Installation planning started
4. RJ31X jack installation planning completed
5. Equipment ordering completed
6. Monitoring information gathering completed
7. Installation planning completed/begin installation
8. Control panel installation completed
9. Control panel preliminary testing completed/all wiring for system completed
10. Installation of protective devices completed
11. Protective devices testing completed
12. Complete operational testing completed
13. Customer instruction on system operation completed

In gathering information for the list, the main concern is usually to make certain that no events are omitted. If PERT is to be effective, all events must be included in the network; failure to report one leads to incorrect schedules and inaccurate resource estimates.

Once the list of events is complete, a network connecting the various tasks in logical sequence can be constructed. Figure 4.2 represents one network that can be developed from the list drawn up for installing an alarm system. It is important to note, however, that this is not the only network or plan that can achieve the objective. Other sequences may work as well or better, depending on the operation of your company.

Event	*Activity*	*Description*	*Time*
1		Contract/proposal completed	
	1→2	Proposal accepted/contract signed	1
2		Preinstallation planning started	
	2→3	Preinstallation planning visit to site	1
3		Installation planning started	
	3→4	Arrange for RJ31X installation	1
4		RJ31X jack installation planning started	
	3→5	Order equipment for job	1
5		Equipment ordering completed	
	3→6	Gather monitoring information	1
6		Monitoring information gathering completed	
	4→7	RJ31X installation	1
	5→7	Equipment in transit	2
	6→7	Set up monitoring account	3
7		Installation planning completed/ begin installation	
	7→8	Install control panel	1
8		Control panel installation completed	
	8→9	Test control panel functions	1
	7→9	Run wire for detection devices and control	1
9		Control panel preliminary testing and wiring completed	
	9→10	Install detection devices and other equipment	2
10		Installation of protective devices and other equipment completed	
	10→11	Test detection devices and other components	1
11		Protective devices testing completed	
	11→12	Perform complete operational test of system	1
12		Operational testing completed	
	12→13	Instruct customer on system operation	1
13		Customer instruction completed/ system installation completed	

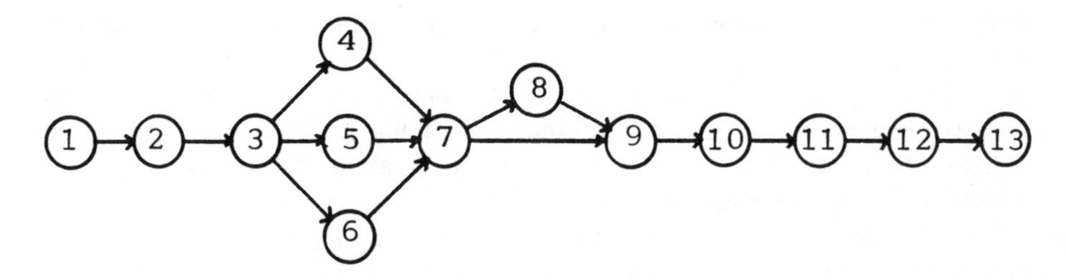

Figure 4.2 PERT network for installing an alarm system. Time (estimated) is given in working days, excluding weekends and holidays.

Network Time Estimates

If the network is to be useful for scheduling purposes, a reasonably good estimate of the time required to perform each activity must be added to the diagram. There is a rather complex formula for determining expected time, one that involves the use of three different time estimates—optimistic time, most likely time, and pessimistic time.

To keep PERT as simple and usable as possible, we will try to estimate the time requirements accurately without using the formula. (If you want more information about PERT and the exact formulas for calculating expected time, college-level business management texts may be consulted.)

As you will note in Figure 4.2, events do not consume time, but activities do. You will also note that the chart shows that several activities can occur at the same time: 3⟶4, 3⟶5, and 3⟶6 as well as 4⟶7, 5⟶7, and 6⟶7. Event 7 cannot be reached until all preceding activities have been completed. It will take one day to have the RJ31X jack installed (4⟶7), two days for the equipment that was ordered to arrive (5⟶7), and three days to set up the monitoring account (6⟶7).

You can estimate the time in any increments that you want: days, hours, or even minutes; calculations using hours or minutes get cumbersome, however. If you use hours, for example, and determine that it will take eighty hours to install a system, you will still need to convert the time to days to schedule your crew. The key to estimating times for the various activities is to be as accurate as possible.

A completed PERT chart provides a great deal of useful information for the project's supervising manager. It enables him to prepare a work-performance schedule listing the expected starting and completion times for each activity. To do this, event 1 (where the project starts) is set at zero on the time axis. The expected time at which each event or milestone will be reached is measured cumulatively from that point. Thus the expected time for any event (circle on the chart) is the largest sum of the expected times on any activity path leading to it. In

Figure 4.2, it would take six days to reach event 7: one day each for activities 1→2, 2→3, and 3→6 and three days for activity 6→7. Event 7 also can be reached by route 1→2→3→4→7 or 1→2→3→5→7. Therefore the most time-consuming path represents the earliest possible time for reaching this milestone because an activity that is initiated by an event, say event 7, cannot begin until all previous activities have been completed.

The expected time determined for the last, or target, event (in this case, event 13) represents the expected overall cycle time required for the installation. In our example, Figure 4.2, it would take 13 days to reach event 13. Suppose, however, that we had agreed to have the installation completed in ten days. We can see that if this PERT network is followed, we will finish the installation three days behind schedule. To install the system in ten days instead of thirteen, we will have to revise our PERT chart and plan the installation differently.

What we have really done so far in this illustration is to determine the critical path for the network, since many paths lead from the initial to the terminal event. The path that requires the longest amount of elapsed time is the critical path.

Critical Path. In Figure 4.2, the critical path is 1→2→3→6→7→8→9→10→11→12→13. The slack along the critical path may be positive, negative, or zero. It will be zero when the completion goal is equal to the length of the critical path. When the completion goal falls on an earlier date than the terminus of the critical path (as in the example), the slack will be negative. Conversely, if the completion goal is further away in time than the terminus of the critical path, the slack will be positive.

Determining the critical path is of crucial importance; it should be the focal point of the analysis, even if it has positive slack. Since the critical path is the series of activities requiring the most time to perform in moving from event 1 to the end of the installation, any event along this route that is delayed will cause the final event, completion of the installation, to slip by the same amount of time.

Implementation of PERT

The actual projects to which PERT is applied are usually more complicated than the example that we have used. A PERT chart for the installation of an alarm system in a large, multistory office building, for example, might require planning for several months and involve hundreds of events and activities. (The network developed to supervise the design of guidance electronics for a guided-missile project was composed of more than 20,000 events.)

Small networks of fewer than 200 events can be set up and analyzed manually. A desk calculator and about eight hours of work are all that are needed. More complex projects should be performed with the aid of a computer.

Summary

The PERT method can be used for any size or any type of project, from planning a ski trip to expanding your office—as well as installing alarm systems. In addition to being a planning tool for managers, it has a secondary function: it is a sales tool. On large projects, if a PERT chart is included with a proposal, your prospect will realize that you have the necessary expertise to plan and execute the job that you are recommending.

A PERT chart displays the sequence and relationship of all significant events in planning how the end objective will be achieved. The relative uncertainty in meeting or accomplishing all activities in the plan is measured and identified.

Presenting a Company Image 5

An image is a concept that is given to the public. Your company's image should create a mental picture that is the sum total of the impressions people have of your business. It is a picture that you deliberately, unwittingly, or accidentally project.

The success of your business depends in part on creating a positive image that customers will notice. This is a continuing process because each individual reaction to your image results in business—or lack of it.

Today, through mass advertising, people purchase a product because of their image of it. For example, jeans were once thought of as work clothes. That image was changed by designers who did little more than add a label to the jeans and raise the price.

In developing an image for your alarm company, your aim is to have it stand out from the crowd of others. Because your competitors are able to supply similar products at similar prices, why should the customer choose yours? If he does, part of the reason will be the effect of your company image.

Image is especially important when selling alarm systems because the customer is not always buying a product he has seen. Instead, he is buying your company—the image in his mind—as well as the system to be installed. In most other businesses, the customer can at least see the merchandise, or something very similar, but when he purchases an alarm system to protect lives and property, he does so largely based on his faith in your image.

Your first tasks are to decide on the type of image you want and determine how to project it. Advertising agencies can help, or you can do it yourself.

DEFINING YOUR IMAGE

As with any task, a definition is needed. You have been asked to define your market; you will be asked to define your goals and objectives. Many other definitions will be required as you organize and manage your business. Right now, take a moment and define your company's image.

Company Name

The name of your company is one very important way of projecting your image, and alarm companies have come up with a variety of names to describe themselves. Just think of some of the more common ones: Reliable Alarms (you can depend on them), Security Engineers (they have the technical ability to do the job), Protection Alarm Company (protection is their business), Protect-Rite Alarms (they do the job right), Custom Alarm Company (tailored to fit), Certified Security Systems (somebody approves of them), and Statewide Alarm Services (they are not small or local). The list could go on and on.

By using adjectives in their names, companies state what they see as their most important quality. Other names can be equally telling. One might designate a location, for example, or it could be a family name. If your company's name reflects the image you want to create, it should be an important part of your advertising.

Company Motto

Mottos seem a bit old fashioned, but they are still useful. If your company's name does not include an adjective, you might use one or more in your motto. Better yet, combine an adjective with an action verb to add emphasis to the image your motto projects.

Make a list of possible adjectives and choose the ones that best describe your operation as you would like the public to perceive it. Include these in a snappy phrase that says what your company does, using as few words as possible in your motto. You might want to emphasize that you are friendly, efficient, or do quality work. You cannot project all of the desirable qualities of your alarm company, so select the one or two that you feel are the most important to your image. A sample list of adjectives includes the following:

Active	General	Prime	Substantial
Alert	Ideal	Protective	Sudden
Bona fide	Leading	Qualified	Superior
Capable	Mobile	Quick	Swift
Custom	Modern	Radiant	Thorough
Diversified	Neighborly	Ready	Ultimate
Dynamic	Paramount	Responsive	Unequaled
Emergency	Practical	Safe	Valiant
Exemplary	Premier	Secure	Vigilant
Fast	Preventive	Smart	Watchful

Logo or Trademark

A logo or trademark can help to create the image you want for your company. You might want one that incorporates the company's whole name or one that uses initials. You might use a design that carries across your company's name or some quality related to it. Animal designs are popular because certain characteristics are associated with different animals. An owl, for example, is supposed to be wise, so a company that wants to have the image of being knowledgeable might use one in its logo.

If you do not have a logo, you can find an artist to design one for you. Your ad agency probably has an artist with whom they work. You could check with local artists' clubs to find one who would do the work free-lance.

If you are creating a logo that will incorporate the name or initials of your company, choose a type style that is appropriate to your chosen image, one that looks modern but is legible. Rather than presenting a puzzle to the customer, you are attempting to have the logo stand out and be easily remembered.

Be Colorful

Another way to have your company be recognized by the potential alarm customer is to choose a color or color combination that will become recognizable as yours. Some colors, like red and yellow, attract the eye. Black lettering on a yellow background creates a special combination. While black and yellow may seem gaudy, the human mind remembers those colors better than any others. Other colors, for example, green and brown, are restful and may not give you the effect that you are seeking.

Use your color or combination of colors every way you can. If appropriate, have colored paper for all of your printed matter: letterhead, booklets, and order sheets. If that is not practical, have your letterhead and logo printed with your color on white paper.

Other Considerations

While your logo, motto, and color will help to define your company graphically and set it apart from the competition, other factors will help define it as well. You can add to your image by displaying photographs of your best installations in your office or showroom and in albums for your sales staff to show potential customers.

Your image is enhanced by your ads, showroom displays, and printed material you send out. Also contributing are the actions and attitudes of your salespeople. Later in this chapter, we take a closer look at how the actions, attitudes, and attire of your company's personnel can contribute to a positive or negative image.

Making a Change

If you decide you want to alter your image, a good time to do so is when you are starting a new advertising campaign. Changing the name of your company is difficult and not advisable, especially if you have been in business for a while and have a following of customers. Updating your logo, however, is an alternative that can be effective.

Once you have decided on an image, a logo, a motto, and a color, use all of these in your advertising, publicity work, and printed materials. They should appear on business cards, stationery, direct-mail materials, and anything else that prospects and customers will see.

COMPANY NEWSLETTERS

Newsletters are an effective though little-used way to project a company image. A newsletter can not only increase communications with your customers, it can also be used to create community goodwill and generate leads for salespeople. The newsletter is a vehicle for communicating with customers and potential customers and tell them about the goods and services that you offer.

An Advertising Medium

A newsletter can include items on security products and special offers, and be used to introduce new products and services. It reminds present customers what you have to offer and it gives prospects an opportunity to learn more about your company.

Dealers who have been using newsletters report that the time, effort, and expense are well spent. Most agree that sales are generated, but it is difficult to determine exactly how many are a direct result of the newsletter. "If we only make one sale from the newsletters we send out," said Doyle Hatcher, president of Protection Alarm Company in Oklahoma City, "they are paid for."

Costs

While costs vary depending on the quantity and quality of the newsletters mailed, a rough estimate is fifty cents per copy, which includes materials, printing, and labor to stuff the envelopes. It does not include postage, though. If the newsletters can be mailed with monthly statements to customers, there is no direct additional postage cost. If they are to be sent out in a separate mailing, add another twenty-five to twenty-eight cents for postage and envelopes.

PUBLIC RELATIONS

Advertising is one method to get your company's name in front of the public; public relations is another. The difference between them is cost. You buy space in a newspaper or magazine or time on a radio or television station when you advertise. Your public relations efforts, though, are more general and are supplied as a public service. There is a cost attached to supplying public relations material, but it is not direct, like paying for space or time.

For example, I donated an hour of my time to an Oklahoma City radio station, KTOK, to be a guest on B.J. Wexler's "Ask the Expert" program and answer listeners' questions about home security. On another occasion, I worked with the local Better Business Bureau, the police department's crime prevention unit, and a local television station to present a half-hour program on residential security.

Both programs involved considerable preparation, to which an economic cost is attached. But I was not buying space or time to promote my business. I was performing a service to help educate the public in crime-prevention techniques. As a result of appearing on these programs, two things happened: first, our company became known as security experts and second, we obtained some new customers because of the exposure.

We might not have been the largest alarm company in Oklahoma City (actually, we were one of the smallest), but we were certainly well known. We presented an image of professionalism, dedication, and public service. Perceiving this image, small size became unimportant to prospective customers; for all they knew we could have been the largest alarm company in town. It simply did not matter.

Be Seen and Heard

To project your company image, you need to be seen and heard. Talk with people at the local newspaper and radio and television stations. Tell them that you will act as an advisor if they should decide to write a story or air a program on security. Make yourself available as a local expert.

You can increase public awareness of what alarm systems can do by meeting and speaking with citizens' clubs, church and business groups, or any other organization that may be interested in security. We have one thing in our favor: crime and crime prevention are uppermost in the minds of many people. All you need is a forum.

Providing Information

When you do send out a news release, be sure that it does not sound like an advertisement. When an editor or a reporter is doing a story, he needs accurate and reliable information, and from a reliable source. He does not need an ad from your company. Unbiased information will give you an edge in being quoted. Find out what information the editor or reporter needs and provide it.

Writing a News Release

The key to a news release is brevity. Also, make it to the point; reporters are trained to look for the "who, what, when, where, and why" behind a story. Tell them.

Send your news release to a specific person, not to "the editor" or to the radio or television station. You can find editors' names listed in the newspaper, and you can call the radio or television station and ask the telephone operator for the news director's name.

Included in the news release should be your name as contact person should more information be required, and the name, address, and telephone number of your company. In addition to including a date on the release, indicate if it is for immediate release or for publication at a later date.

At the top of the release should be a one-line summary of what you are saying, for example, "Captures at the Scene Increase when Alarm Systems Are Used." Then proceed with a short, factual story.

Send a cover letter with your news release stating what your company does and that you are sending the information as a public service in the interest of preventing crime (or fire, depending on the story).

A few days after you send the release, follow up with a telephone call. Ask if the information was sufficient or if additional facts are needed. Also, ask what type of item is generally used by the newspaper (or radio or television station). It will give you a few ideas for your next news release.

Not everything you send will be put into print or on the air. Be patient and persistent.

Disaster Planning. "The best laid schemes o' mice an' men . . ." are all subject to Murphy's law (my apology to Robert Burns). Even with the best of public relations programs, things can go wrong. For example, you have an interview with a local reporter. After you have explained the benefits of having an alarm system, he asks you about his neighbor's alarm that goes off every morning at 3 A.M., or a business that had an alarm system that did not work and suffered a loss.

Your best approach is honesty and forthrightness. Answer the questions simply and briefly, being as factual as you can. Do not speculate about causes or attempt to place blame. There are problems in every profession. Ours is no different. We just have to try harder to overcome the negative images and create positive ones.

PROFESSIONALISM

Over the years, I have heard the phrase "don't judge a book by its cover" innumerable times. Think about it for a minute. Isn't it the cover that prompts you to pick up the book and investigate further to see if you want to read it? Isn't the same basic philosophy applicable to the people and companies with whom we do business? Right or wrong, you and your employees, and eventually your company, are judged by the impression—the image—that is reflected in your attitude, action, and attire.

Attitude

Attitude is important. Pleasant, helpful salespersons yield higher results than those who are unfriendly or preoccupied. The importance of a good attitude does not apply solely to sales staff, however. Every member of the organization is responsible for projecting a positive company image.

The staff member with whom a customer or prospect is currently dealing does, at that particular time, represent the entire company. Whether it is a rushed secretary on the telephone or an installer who brushes off a customer's question, the image of the company is resting with that employee.

A good attitude does not require much. A smile or a friendly greeting is sometimes all that is necessary.

Action

Certain words convey the meaning of action: dynamic, effervescent, energetic, and aggressive are just a few. The whole idea behind being a company of action is to get the job done. An important part of your reputation, again, your image, will be your reputation for taking action. We are in a problem-solving business. Protecting lives and property with our products and services is simply a solution to our customers' problems. Generally, those problems are rooted in one basic problem: the need to effect a safe end secure environment.

Whether we are following up on an installation, making a sales call, or dispatching service personnel, we need to present the image of an active, dynamic, effervescent, energetic, and aggressive alarm company.

Attire

Before you can sell your product, service, or idea, you have to sell yourself. Selling begins with the prospect's first impression of you (or one of your staff members). If you are fully prepared for the sales call, the impression will be a favorable one.

Because different people have different tastes in clothes, it would be impossible to tailor your attire to each prospect's preferences. A general rule is to be conservative.

Whether you are calling on an executive in his office, a homeowner in his home, or an auto mechanic in his shop, a simple, conservative suit will meet most requirements. The mechanic may think that you are a bit overdressed for his greasy shop or the style-conscious prospect may think that you are not a fashion plate; both, however, will probably agree that you are well dressed and look like a professional. A well-groomed appearance that includes neatly styled hair down to freshly polished shoes is important in creating a favorable first impression.

According to psychologists, a positive first impression lasts a long time, and so does a negative one. You will have enough obstacles to overcome without adding one that is not completely under your control. Thus appearance is important for all staff members, not just the sales staff. While your installers may be top-notch professionals, they could be perceived as careless and sloppy if that is the way they dress. Even if they do not wear uniforms, they should wear traditional work clothing. Jeans and work shirts will suffice if they are clean and neat.

The same basic concept of having a good, professional appearance should be followed by the office staff. Whether it is your secretary or your inventory clerk, if they meet the public, they should project your company image.

6 Operating the Business

The title of this chapter is rather general. Obviously, we cannot possibly discuss every conceivable aspect of operating your alarm company. The topics that we do consider are those that will have a significant effect on the bottom line—your company's profits.

In Chapter 2 we discussed some of the capital requirements for starting your business. Whether you are just starting or have a continuing operation, fiscal matters deserve close attention. Budgeting is important, so is measuring your success. Both require accurate records.

FINANCES

Most owners and managers of larger alarm companies understand the importance of financial planning. Unfortunately, many owners and managers of smaller alarm companies do not. This may, in fact, be a major factor in the high attrition rate for the small companies. It certainly appears to be what separates the business-oriented and financially successful small concerns from the ones that are struggling to pay their bills. The most common failing is not keeping a good set of records.

"For the individual just going into business, experience clearly shows that an adequate recordkeeping system helps increase the chances of survival and reduces the probability of early failure. Similarly, for the established business owner, it has been clearly demonstrated that a good recordkeeping system increases the chances of staying in business and of earning large profits."[1]

There are several reasons why every alarm dealer should have a record-keeping system especially attuned to his individual operation. The most obvious is that it helps him to survive in business. It is in your self-interest to understand the financial workings of your company. As an alarm dealer (owner or manager), you will be responsible for many of the things that your former employer arranged for you. Health insurance, life insurance, and a pension plan are often part of a benefits package supplied by the employer at little or no expense of time or money to you when you were an employee. As a dealer, you are now responsible.

You should be concerned with three areas of financial planning. The first is setting up a basic records system so you can keep track of what you spend and what you earn. The second area is budgeting, so you can tell how you are doing and whether or not you are going to survive. The third area is planning a benefits package that protects you personally, your employees, and also your business.

Record Keeping

A record-keeping system lets you keep a close check on how you are doing financially—what you are earning, how much you have spent, who owes you money,

[1] Small Business Bibliography no. 15, "Recordkeeping Systems—Small Stores and Service Trade," Small Business Administration, Washington, D.C.

who has paid you, and whether they have paid you the correct amount. Even if you have an accountant, which is recommended, you still need to understand enough about the financial aspects of your company to talk intelligently with him. Also, you will need accurate records if you plan to obtain any kind of financing for your business, especially a bank loan. The last reason that you need a records system concerns the IRS. It is through your business records that you may be able to justify any expenses that you deduct. The IRS has no set requirements, except that you must maintain permanent records that can be used to identify income and expenses. The burden of proof, should you be subjected to an audit, is on you.

Budgeting

Once you have established a records system, the next step is to draw up a budget. A budget is the basis of a record-keeping system. It can be prepared for any length of time—a month, a year or longer. Many small-business owners, including alarm dealers, think that they do not need to prepare a budget, but they do. Every time you start worrying about money, usually in the middle of the night, and finally sit down and add up what is coming in and what has to go out, you have, in effect, prepared a budget. You will be much better off by doing this on a regular basis and filing budgets for reference. It is much simpler than panicking.

A budget is a planning tool. It helps you set a goal and move toward it. It shows whether or not you are making a profit. Ideally, to do a budget you should know what your income will be, something that alarm dealers cannot always easily predict. Having a predictable income, recurring revenue from leases, and monitoring helps. After a while, you will begin to have an idea of what your yearly income will be.

There are many kinds of budgets. For our discussion, we will keep it simple.

	First Month	*Second Month*	*Third Month*
Beginning cash balance			
Income			
Cash payments			
Rent			
Utilities			
Operating costs			
Salaries			
Total net income before taxes			
Estimated taxes			
Balance			

Figure 6.1 Format for a simple, quarter-year, monthly budget.

Item	*Budget This Month*	*Actual This Month*	*Variation This Month*	*Budget Year to Date*	*Actual Year to Date*
Income					
Sales					
Leases					
Service					
Monitoring					
Expenses					
Salaries					
Inventory					
General/administrative					
Sales expense					
Transportation					
Supplies					
Net income					

Figure 6.2 Format for a budget showing actual and projected expenses.

After you have an understanding of the basics, you can proceed into the more complicated budgets.

Figure 6.1 shows the format for a very simple budget. Using this format, you can tell whether or not you will be able to meet your monthly expenses. Another type (Figure 6.2) shows your actual and projected expenses. It will indicate whether you are adhering to the budget that you have established.

The budget forms in Figures 6.1 and 6.2 are very simple. Obviously, there are more items that you, as an alarm dealer, will want to include in yours.

SIMPLE ANALYSIS

Comparing current data to previous data is a simple method to measure success. How (and what) is your company doing this year compared to last year? The time periods used will depend on your particular operation. It might be more beneficial to use smaller units; for example, what was your sales volume for this March compared to last March? If you want more specific information, you can make comparisons on a week-by-week basis.

Comparative Data

Sales volume need not be the only category. You may want to compare other factors to obtain an overall picture of your company's progress.

Listed below are some questions. The answers you give will provide information that can be analyzed.

On a sheet of paper, make three columns and list these headings: (1) to be compared, (2) this year, and (3) last year. (Of course, you could change "year" to a different time period. For example, your chart could be for "this March" and "last March.") On your chart, fill in the information for columns 2 and 3 as you answer the questions below, which are for column 1.

- Total residential customers at end of period?
- Total commercial customers at end of period?
- Total customers (residential and commercial) at end of period?
- New residential customers acquired during period?
- New commercial customers acquired during period?
- Total new customers (residential and commercial) acquired during period?
- Total loud/local systems at end of period?
- Total monitored systems at end of period?
- New loud/local systems installed during period?
- New monitored systems installed during period?

These are only a few of the questions that could be asked to generate data for comparisons. Others might be about service calls or the number of systems upgraded during the period. You could be even more specific and ask questions about types of systems: burglary, hold-up, medical, and fire. You could determine the average sales for different types of systems, for example, for a residential local burglar alarm or commercial monitored burglar and fire system. All of these data can be derived from information that you should already have available.

To make our chart more useful, we could add two more columns: (4) gain or loss and (5) percentage of change. To obtain the data for column (4), subtract the amount in column 3 from column 2. Dividing the amount in column 4 by the amount in column 3 will give you a figure that represents percentage of change, for column 5. Table 6.1 shows an example of a simple analysis.

Table 6.1 Example of a Comparative Analysis for ABC Alarm Company

To Be Compared (*1*)	*This Year* (*2*)	*Last Year* (*3*)	*Gain or Loss* (*4*)	*Percentage of Change* (*5*)
New residential customers for year	50	40	+10	0.25 (+25)
New commercial customers for year	40	35	+ 5	0.14 (+14)
Total new customers for year	90	75	+15	0.20 (+20)
New monitored accounts for year	75	55	+20	0.36 (+36)

Analyzing specific parts of your income-producing operations may pinpoint activities that are more or less profitable than others. It can serve as a method for spotting problem areas. If you had predicted a 25% increase in residential sales and only achieved a 21% increase, are there specific reasons that your goal was not achieved? Conversely, if you achieved a 40% growth in residential sales, can you determine why your estimate was too low? Could it be that while residential sales rose, commercial sales dropped?

BEAT YESTERDAY

Another method for tracking trends and comparing current operations with past performance is through the use of a "Beat Yesterday" book that charts present sales or installations on a monthly basis. Comparing data for small time periods may be helpful in determining trends or cycles in business and will assist you in forecasting and budgeting. A sample page from a Beat Yesterday book might look like the one shown in Table 6.2.

The Beat Yesterday book should have a space for notes. Using the example shown in Table 6.2, we can observe several trends. These and any other explanatory information should be included in the notes at the bottom of the page.

In the example, we can see that residential installations decreased during March of year 3, possibly because commercial installations doubled from the previous year. Total installations, residential and commercial, increased in year 2 and year 3, but remained constant for year 4. An explanatory note would help clarify the reason. Maybe the lack of growth was caused by the economy or maybe it was caused by the weather (an extremely stormy March).

A Beat Yesterday book is simple to create and use. It can contain whatever

Table 6.2 Sample Page from "Beat Yesterday" Book for ABC Alarm Company for the Month of March

Comparison of Installations	*Year 1*	*Year 2*	*Year 3*	*Year 4*	*Year 5*
Residential	7	9	8	10	
Local/loud	6	5	5	4	
Monitored	1	4	3	6	
Commercial	4	5	10	8	
Local/loud	2	3	6	4	
Monitored	2	2	4	4	
Residential	7	9	8	10	
Leased	3	4	4	6	
Sold	4	5	4	4	
Commercial	4	5	10	8	
Leased	3	4	8	6	
Sold	1	1	2	2	

information that you think will be useful to you in planning for the growth of your business.

ACCOUNTING

Business may be good. Sales may be up. But are you making a profit? Increased revenue does not necessarily mean increased profit.

Accounting System Components

Just like an alarm system, an accounting system has components. The two most important ones are the balance sheet and income (profit and loss) statements. The basic accounting formula used to prepare a balance sheet is: assets = liabilities + net worth.

Assets are the things owned by the company such as cash, equipment, supplies, and inventory. Liabilities represent the company's financial obligations, what it owes to others. Net worth, sometimes called owner's equity, is what belongs to the owner (or the business) after the liabilities have been paid.

To prepare a simple balance sheet, assets, liabilities, and net worth are listed as of a certain date. It is often desirable to prepare a monthly balance sheet that summarizes the status of the company. Figure 6.3 shows ABC Alarm Company's balance sheet for June. You will notice that assets include cash (in the bank and on hand), accounts receivable, inventory, supplies (installation, service, and office supplies), furniture and fixtures, and equipment (vehicles and office equipment). Also included under assets are buildings and land, which are considered fixed assets.

The liabilities section includes accounts payable, notes payable, and taxes payable. Accounts payable are usually short-term obligations, such as credit accounts with distributors; notes payable are long-term obligations, such as bank loans. Taxes payable may include payroll, income, or property taxes that have accrued but have not been paid.

Another accounting system component is the income statement, often referred to as a profit-and-loss (or P and L) statement. The income statement shows the details of the company's operations over a specific period, usually a month.

The two primary sections of the income statement are revenue and expense. Revenue reflects the gross income of the business from the sale of alarm systems or service. If you lease equipment, it will include the income from leases. If you offer monitoring services, that income will be included. The various categories for revenue will depend on your particular operations.

Expenses are those funds spent by your company to do business. Salaries, utilities, supplies, and transportation are a few of the expenses that will be shown on your income statement.

Assets		
Cash		
Cash on hand	$ 270.00	
Cash in bank	3,640.00	
Total cash		$ 3,910.00
Accounts receivable		1,675.00
Inventory		2,780.00
Supplies		
Installation	705.00	
Service	340.00	
Office	260.00	
Total supplies		1,305.00
Furniture and fixtures		4,790.00
Equipment		
Vehicles	9,680.00	
Office	3,645.00	
Total equipment		13,325.00
Buildings		19,800.00
Land		14,670.00
Total assets		$62,255.00
Liabilities		
Accounts payable		$ 2,500.00
Notes payable		34,900.00
Taxes payable		3,850.00
Total liabilities		$41,250.00
Net worth		
John Smith, capital		$21,005.00
Total liabilities and net worth		$62,255.00

Figure 6.3 Sample balance sheet for one month.

The basic formula for an income statement is:

$$\text{revenue} - \text{expenses} = \text{net income (or loss)}.$$

Figure 6.4 shows ABC Alarm Company's income statement for July.

The balance sheet and the income statement in the examples are simplified for the purpose of illustration. The actual format and the amount of information included in your company's financial reports will depend on your operations. It is usually better to have too much information instead of too little. Your accountant will be able to guide you in deciding on the format and type of information that will be most useful to you.

Return on Investment

One way to measure growth is to compare this year's gross revenues with last year's. It will indicate how well your company has done at producing income.

Revenues		
Sales	$3,230.00	
Leases	4,100.00	
Service	1,110.00	
Monitoring	1,760.00	
Total Revenues		$10,200.00
Expenses		
Salaries	$4,100.00	
Inventory	1,360.00	
General/administrative	1,260.00	
Sales expense	1,210.00	
Transportation	620.00	
Supplies	330.00	
Total expenses		$ 8,880.00
Net income		$ 1,320.00

Figure 6.4 Sample income statement for one month.

If your income (gross revenue) was $100,000 this year and $80,000 last year, you increased your income by $20,000. Dividing the increase ($20,000) by last year's income ($80,000) yields the percentage of increase: $20,000 ÷ $80,000 = 0.25 (or, 25%).

It is possible, though, for gross revenues to show an increase and for profits to decrease or for a loss to occur. If your costs (expenses) increased significantly during the year, a true picture of your financial condition would not be presented by looking at gross revenues alone. After all, it is the profit figure that is important to successful operations.

If your income for this year was $100,000 and your expenses were $90,000, you showed a profit of $10,000. If the profit for last year was $9,000, you only increased your profit by 11%, not 25% as you might have thought from the revenue figures.

Dollar amounts for revenues, expenses, and profits are useful figures. They become more useful, however, when combined with the figures mentioned earlier: How many residential alarms were installed this year compared with last year? How many were monitored? And so forth.

Your gross revenues may be up because of sales instead of leases, which provide more capital initially but may cause problems later because leasing would have produced a recurring income. Or maybe monitoring income is up, which increases the gross revenue figure, but actual installations are down.

While profit is the primary objective, there is an important point to consider: how much of an investment was required to make the profit? Investing $100,000 to make $1,000 profit might not be worth while; yet making $10,000 profit with the same investment is much more attractive. Measuring your company's success based on the relationship between what you have invested and the income that you derive from that investment is a significant evaluation. This type of analysis

Total revenues = $100,000
Total assets = $20,000
Profits = $25,000
Asset turnover = total revenues ÷ total assets
Asset turnover = 5.0
Profit-to-sales ratio = profits ÷ total revenues
Profit-to-sales ratio = .25
Return on investment = asset turnover × profit-to-sales ratio
Return on investment = 1.25 (125%)

Figure 6.5 Example of how to calculate ROI.

is called return on investment (ROI). It is calculated by using asset turnover and profit-to-sales ratio. The basic formulas are shown below:

Asset turnover = total revenues ÷ total assets.
Profit-to-sales ratio = profits ÷ revenues.

The figures needed for the formulas can be obtained from your current balance sheet and income statement.

Multiply asset turnover by profit-to-sales ratio. The result is your company's return on investment. Expressed as a formula, it would look like this:

ROI = asset turnover × profit-to-sales ratio.

Figure 6.5 shows an example of calculating ROI.

Now that we know ABC's ROI from Figure 6.5, what does it mean? Fortunately, ABC has an excellent ROI. The 1.25 figure means a 125% return on investment. Had the ROI been lower, we might have wanted to analyze it more closely and compare it to various interest rates for savings accounts, certificates of deposits, or treasury bills, for example. If we could obtain a better return on our investment from risk free investments, we should seriously analyze our operations to plan for a higher ROI. As another example, if ABC was borrowing money from a bank at 18% interest and only generating a 15% ROI, it would be falling behind rather than gaining.

Inflation and ROI

If the inflation rate is the same as or higher than your ROI, you are not making any real progress. Any returns that you generate are being eroded by the declining value of the dollar.

Keeping your company's ROI at an acceptable level requires careful planning and cost control in addition to building sales revenues. While there are numerous variables to consider and control, two that can have a significant effect on profits are inventory control and credit (and collection) management.

INVENTORY CONTROL

Whether your operation is large or small, the quantity of inventory that you carry can affect your profits. Small dealers may find it more difficult to spend time on inventory control, whereas larger alarm companies may have a full-time staff member assigned to that function.

Too Much

Having too many items in stock can be costly. There is more involved than the money paid for the inventory. A few of the costs involved in maintaining an inventory are record keeping, labor, shipping charges, obsolescence, storage, and taxes.

When items are received, record-keeping entries must be made. The items must be checked off the purchase order, the invoice must be checked against the purchase order, and a check must be issued; all of which takes time. Moreover, it takes time for someone physically to receive the shipment, unpack it, and place the merchandise on the shelves.

Every time a shipment arrives, shipping charges are incurred, unless the shipper pays the freight. The shipment may arrive with freight collect or the charges may be added to the invoice. Either way, it is an expense.

Equipment that is ordered in large quantities may become obsolete if not used quickly. A stock room full of dated equipment, even if it was purchased at a low price, is no bargain.

There is a cost attached to storing equipment, especially if you have to rent extra space. If your state taxes inventories, you will have a larger tax bill.

A conservative estimate for the cost of keeping an inventory is close to 20%. That is, for every dollar of inventory (average, on a yearly basis) it will cost you twenty cents per year.

Too Little

It is difficult to attach a cost to having too little inventory. If you do not have enough equipment to install a system, however, you may lose the sale or postpone getting paid.

Sources of Supply

Part of the answer to the inventory control problem may be in selecting a supplier. If he has a good inventory and shipping time is minimal, your inventory control problem can be minimized.

A distributor who carries a variety of products can be particularly valuable to alarm dealers. In effect, a well-stocked distributor's warehouse becomes an extension of the dealer's stock room. Most distributors also offer more than a wide selection of products, providing technical and application assistance as well.

Of course, equipment may be purchased directly from manufacturers. It is usually simpler, though, to purchase from a distributor, especially for small quantities. Distributors' prices are comparable to manufacturers' and shipping time usually is less.

SELECTING A DISTRIBUTOR[2]

The current wholesale alarm equipment market benefits the alarm dealer by providing him easy access to over 200 manufacturers and more than 100 distributors—the majority of whom market nationally. Most manufacturers and distributors offer toll-free telephone lines for ordering. Access to equipment, therefore, is only a phone call away, regardless of the dealer's location.

Distributorships are relatively new to the alarm industry, emerging in the early 1970s and growing steadily ever since. Alarm equipment distributors may be generally classified by type of operation.

First, are the supermarket-type operations that offer extensive selections. Dealers who know exactly what they want usually find the equipment readily available. Application and installation assistance may be available on a limited basis.

Second, are the discount-house operations. Relying on a quick inventory turnover from national marketing efforts, this type of distributor may offer the dealer lower prices in exchange for other services, such as technical assistance and equipment return privileges.

Third, are the full-service distributors. Often, they stock specific lines of top quality equipment, in addition to providing installation, application and general technical assistance to the dealer. If requested, many full-service distributors will recommend specific types of equipment and assist the dealer with the selection of equipment for specialized applications.

The primary advantage to purchasing from a distributor rather than a manufacturer is convenience. The distributor offers a complete line of alarm equipment and installation hardware, whereas the manufacturer may not. A local distributor may be able to ship the products more quickly, with many of them shipping the equipment the same day that it is ordered. Since most distributors offer equipment at "factory-direct" prices, there may not be a price advantage to purchasing from the manufacturer.

Checklist for Selecting a Distributor

1. Is name brand, quality alarm equipment in stock?
2. Are sales personnel familiar with the equipment and know what additional equipment (such as brackets, transformers and batteries) may be needed?
3. Is installation, application and general technical assistance available?

[2] Joe Lanier, Defensive Security Southwest, Inc., Jasper, Texas.

4. Are technical seminars and trade shows sponsored by the distributor?
5. Is a complete catalog available?
6. Is "factory-direct" pricing offered? Quantity discounts?
7. Are discounts offered for pre-payment, COD or payment of invoices within a specified time?
8. Is credit available to qualified dealers?
9. Is there a toll-free ("800") telephone number available for placing orders?
10. Are orders usually shipped within 24 hours?
11. Does the distributor have a good relationship with local dealers?
12. Are local, regional and national dealers' associations supported by the distributor?

CREDIT AND COLLECTIONS

Unless your business operations are totally different from those of most alarm companies, people will owe you money. Extending credit is a business fact of life, especially if you lease systems.

Once you install a system, unless it is paid for in advance, your customer is obligated to you. He has the equipment, you have his signature on a promise to pay. If his promise is good, there will be no problems; if it is not, it will be expensive.

Even though you have a good contract and promissory note, trying to collect a delinquent account will require the time and effort of someone on your staff. If the matter requires litigation, you will have an attorney to pay.

A less obvious cost of bad debts is seen in your business' cash flow. If you cannot collect from your customers and cash flow is restricted, you may have to apply for a bank loan for working capital. Added to the other collection expenses will be interest on the loan.

Again, accurate record keeping is important. You should know how many of your customers owe you money and how much they owe you. As a general rule, less than 15% of your total receivables should be more than sixty days old. Try to avoid letting any accounts become delinquent for more than 120 days. The older the account, the more difficult it is to collect.

As soon as an account becomes overdue, you should begin efforts to collect. If a customer has thirty days to pay his account and it is not paid on the thirtieth day, you should start your collection process by sending a late payment notice.

The notice should be sent on the day that the account was to have been paid. It should take the form of a reminder and say something like this: "SECOND REQUEST FOR PAYMENT. We were happy to extend credit to you and would appreciate your check by return mail."

Ten days later, if payment still has not been received another notice should be mailed. This one should say: "OVERDUE ACCOUNT. This account is overdue. Please mail your check at once!"

If, after another ten days, the account is still unpaid, a final notice is sent. Wording should be similar to the following: "FINAL NOTICE. We cannot wait any longer. Avoid trouble and expense by sending your check at once!"

About 75% of delinquent accounts will be paid after mailing the second notice. Most of the rest will be paid after sending the overdue account notice. Some will respond to the final notice and a few will require additional effort before collection can be made.

Accurate records of mailings are a must. Should you need to pursue legal action, you will have to show the court the efforts that you have made to collect.

If collection has not been effected by the time the overdue account notice is sent, personal contact is suggested, either by phone or in person, and regular contact should be maintained until the account is current. Consumer protection laws restrict the frequency and times that personal contact may be made. Be sure to check local laws before attempting to collect from difficult nonpaying customers.

Although collection agencies charge from 25% to 50% of the total amount due, it may be the only way to recover part of the revenue.

PUT IT IN WRITING

A good friend of mine asked about having an alarm system installed, so we met over lunch to discuss it. Since I was familiar with his home, I had already designed the system and was prepared to explain my suggestions. During lunch, I took a seven-page proposal/agreement from my briefcase to show my friend the equipment that I was recommending, the services that my company would perform, and what it would cost, in addition to the standard clauses contained in our contract. I proceeded to say that I would insist on such an agreement even if he were my father. "I think of this as a way of protecting not only our business relationship, but our personal relationship," I explained.

All of the details of the system and its cost were included in the proposal/agreement. I knew what the agreement said and, after reading it, so did my friend. Later, neither of us could claim that there had been a misunderstanding.

When someone who is not your friend is reluctant to agree to a project in writing, be wary. Ask yourself why. There could be a variety of reasons; few of them will be to your benefit.

Occasionally, someone will resist putting something in writing because that just is not the way he does things. This is still no reason for you not to do so. A closer look at contracts and liability is presented in Chapter 19, which will show you why it is extremely important to have a written agreement.

Documents other than contracts should be in writing, too; invoices, statements, and purchase orders come to mind immediately, and there are many others for our type of business. Keep one thing in mind: should a problem arise, supportive documents will be invaluable.

Documents and the IRS

Another important reason for putting things in writing is to satisfy the Internal Revenue Service. Without the support of written agreements, invoices, and other documents, you may have trouble proving what your taxable income is. If you are thinking that could be to your advantage, you may be in for a surprise. The IRS has methods of "reconstructing" income data. Unfortunately for the small-business owner, it often means a larger tax liability than he would have had if he had been keeping accurate records.

SMALL BUSINESSES AND THE IRS

Sometimes IRS agents take a rather dim view of small businesses, especially sole proprietorships, because they think it might be easy to hide income and evade taxes. Right or wrong, having to spend the time explaining your income and tax deductions to the IRS is a very nonproductive way to spend your time.

Regardless of the type of organization you have—sole proprietorship, partnership or corporation—you must pay taxes quarterly. Failure to do so eventually will cause a problem and you will have to pay interest and penalties. Your money can be better spent on other things.

On the one hand, the IRS has the power to check your accounting records and bank accounts. They can even obtain a court order to enter your safe deposit box. There is not much you can hide from them, nor should you try. On the other hand, they are not out to get you. After all, the more your business prospers, the more taxes you pay.

Quarterly Tax Payments

All businesses, and some individuals, are required to pay income taxes on a quarterly basis. This aspect of paying taxes seems to bother small businesses more than large ones for some reason, yet it is a relatively simple procedure. You estimate your tax bill for the entire year, divide by four, and make those payments. Unless you are on a different tax year (fiscal year, as opposed to a calendar year), on April 15, like everyone else, you figure out what you have actually made and either pay the difference or file a claim for a refund. Quarterly income tax payments are due on April 15 (when you also owe the unpaid amount on your previous year's tax bill), June 15, September 15, and January 15.

Although the IRS rules are subject to change, currently you must pay 80% of your annual tax bill through the quarterly payments or risk a 20% penalty for underpayment. From this point, the rules become somewhat complicated. A good accountant is an asset.

Social Security Taxes

Depending on your organizational structure and whether or not you are paying yourself a salary, social security taxes may pose another tax problem. If you are receiving a regular salary, taxes should be withheld from your earnings. If, however, you are withdrawing capital instead of taking a salary, you will have to pay social security taxes at the end of the year. Based on your net income (capital withdrawals), you figure how much social security tax you owe. The IRS Schedule SE is used for this purpose.

Tax Assistance

In addition to, or in lieu of, your accountant, the IRS does provide answers to tax questions. If you have a simple question that can be handled over the phone, they provide free advice.

There is a major problem with taking advice from the IRS: their advice is not binding. You may call five different agents and get five different answers. The argument that "the IRS agent told me" just does not hold water. Like most other government agencies, they have a built-in hold harmless clause; nothing is binding, even if it is in writing. There are exceptions, of course, but who needs the aggravation? Hire a good accountant.

COMPENSATION AND BENEFITS

Determining wages and benefits is not an easy task. If you pay too much, you will be wasting money and reducing profits; too little and you may not be able to retain your employees.

There is no universal wage scale. Many times, there are no good guides of any type. Table 6.3 shows some suggested hourly rates for various nonmanagement positions. The table should be used as a guide, it is not an absolute scale. Many factors influence local wages.

Table 6.3 Suggested Hourly Wage Ranges for Various Nonmanagement Positions

Position	*Average Employee ($)*	*Good Employee ($)*
Installer	6.50–10.00	9.00–17.00
Installer's helper	5.00–7.50	7.00–10.00
Installer trainee	5.00–6.00	5.50–6.50
Troubleshooter	10.00–15.00	12.00–17.00
Technician	12.00–16.00	15.00–20.00
Secretary	5.50–7.00	6.50–9.50
Clerk-typist	4.50–6.00	5.00–7.50

A benefit package is a necessity in today's economy. Employees are looking for compensation and fringe benefits. In addition to health, major medical, and life insurance, employees seek pension and profit-sharing programs. To develop an economical yet attractive benefit package, you will probably need to meet with your accountant, insurance agent, and banker. It is not unusual to spend as much as 30% of salaries on a benefit package, though most alarm companies spend about 15 to 20%.

"YOUR COMPANY GAME PLAN"[3]

Ronald I. Davis

This article was written within a few hours of completing our Spring grouping of management and sales seminars around the country where probably three to four hundred companies were represented. Because the seminars were spread out over three days, I had an unusual opportunity to talk with owners and managers of alarm companies about the problems they are encountering in proper administration of their organizations. These problems were manifested in many ways: poor return on investment, high turnover of personnel, poor controls over inventory and rolling stock, poor company image within the community and overall management inefficiencies.

This is not meant to be a scathing indictment of alarm companies, but rather a statement of awareness as to what some of the problems are. In most cases, the presidents and managers of companies represented at the seminars indicated that their companies have experienced such rapid growth that they were unable to maintain adequate control. That's the old "good news, bad news" dilemma that managers face. The good news is rapid growth; the bad news is poor control.

Management by objectives (MBO) is a phrase that is frequently used by consultants, management experts and top level managers in large companies to describe a program that leads to predetermined results. Those results are planned for ahead of time, are of a positive nature, and all employees and executives within a company are geared to those goal accomplishments. However, partially because of the rapid growth experienced within the alarm industry, many alarm company owners and managers have not implemented management by objective programs, nor have they even developed the tools which are necessary to implement an MBO program.

Several of those tools, the company manual and the personnel policy manual, are essential for a proper understanding of the "ground rules" of running a successful business. MBO and more or less sophisticated goal orientation programs aside, the development of a personnel policy manual is important just from the thought processes that take place in their development and subsequent reviews. These manuals provide a checklist for anyone in our industry who wants to be able to "track" the progress of his or her company against some predetermined criteria for growth.

---

[3] Reprinted with permission from *Alarm Signal,* National Burglar and Fire Alarm Association, July/August, 1980.

Ronald I. Davis is president of Davis Marketing Group, an international consulting firm specializing in the alarm industry, as well as president of the Security Alliance Corporation, a national organization of franchise dealers specializing in alarm products and services.

Developing a Company Manual

The primary purpose of a company manual may be somewhat different from the results you are seeking to achieve from that manual. The results are an opportunity to develop a predetermined roadmap. That's the difference between management by objectives and management by crisis.

With a roadmap, you can establish worthwhile, predetermined goals. Management by crisis is a phrase used to describe managers who react to situations and make decisions *after* a problem has occurred rather than think it out ahead of time. But one of the great advantages in development of a company manual is that it causes you to think about some things you may not normally take the time or effort to contemplate.

Listed below is a checklist of items you might want to incorporate in a manual. Remember, it's not critical that this company manual be lengthy, expensive or a grammatically perfect corporate document. Rather, it should reflect your thinking in simple language and be easily understood. Please make a note of those items for which you want to develop a "statement," and then refer later to this list as you prepare the manual:

1. Introduction—A statement as to why you're writing the manual, who should be reading it and under what circumstances it should be read (i.e., new employees for orientation, people who are promoted within your company, mandatory reading once a year, etc.)
   - A. Brief company history—Very frequently new employees don't have an understanding of the history of your organization, its growth and background. This should correct that situation.
   - B. Company philosophy—What is your company's philosophy? What business are you in? Is the primary goal to increase monitoring? Service? New installations? Develop new product lines? You should sum up in a few sentences what your company stands for.
2. Organization of company
   - A. Explanation of chain of command—Who reports to whom? Why? When? Who has responsibility for what? And do each of the people who have management responsibility have authority as well as responsibility for getting a particular task done?
   - B. Organizational chart—This is nothing more than a visual statement of relative importance and direction of people within your organization. Constructing this chart might be helpful to you in making decisions about where people belong within the framework of the company.
3. Overall operating policies—What are the general operating policies of your company? This would relate to many of the things covered in a personnel policy manual which will be addressed later on in this article.
4. Overall personnel policies—This too will be more completely developed under the heading of personnel policy manual.
5. Overall purchasing policies—Who has authority and responsibility for purchasing products, making capital expenditures, developing a tracking system for purchase orders, invoices, etc.?
6. Housekeeping, maintenance and company security—What are the rules and regulations regarding basic maintenance and cleanliness of your operation? Who

has that responsibility and how does each employee fit into that overall picture? What about company security? You should set up policies that eliminate the chances of compromising your company's security.

7. Supplies and inventory—Who has the responsibility for basic products inventory, service equipment inventory, office equipment inventory and office supplies inventory? Do you have predetermined levels of reorder? If so, who has to approve those levels?
8. Forms, systems, filing procedures, update and review of accounts and systems procedures—This section should be utilized for a listing of all of your corporate forms, filing procedures and internal controls. Each form should be identified in the manual, along with explanations of how it's used, under what circumstances, etc.
9. Departments and functions—Even small companies have to departmentalize their activities. Some of the divisions you might need to develop a function sheet for are:
   A. Accounting and administration
   B. Sales and marketing
   C. Purchasing and inventory control
   D. Manufacturing (if applicable)
   E. Customer service
   F. Central station administration
   G. Investigations
   H. Guard service accounts
   I. Patrol accounts
   J. Lie detector and personnel verification
   K. Personnel
   L. Engineering
   M. Research and development
   N. Quality control
10. Methods for upgrading manual, as well as systems and procedures for upgrading of all company operations—This section would identify how employee suggestions are to be utilized (and financial incentives for profitable ideas). Implementation and distribution of the policy manual—who is to receive copies and what is their expected response. Finally, who has the control of the manual to make sure it is utilized and that there are no questions as to its availability.
11. A detailed index—This should be developed after the manual is written, listing headings and subheadings with their page numbers.

Developing a Personnel Policy Manual

Of equal importance is the personnel policy manual, which is a companion to the company manual. Through the development of a personnel policy manual, you will have established, prior to emergence of any crisis, a detailed, point-by-point statement of the operating policies of your company with regard to personnel. The following listing is intended only as "idea stimulators" for you to consider in the development of your own manual.

In lieu of a manual, you might want to issue separate memos or bulletins regarding some of the subjects listed below. A complete manual will encompass the majority of these points.

Employee Relations

1. Supervisors' responsibilities to: the company, immediate supervisor, other supervisors and subordinates.
2. Supervisors' responsibilities for cost control and paperwork.
3. Employee relations program.
4. Company position toward union affiliation and union bargaining representative.

Hours of Work

1. Definition of work day and work week.
2. Define hours of work which are compensable.
3. Define shift operations (if applicable).
4. Policies on hours of work compensable at straight time and at overtime premium.
5. Policy on maximum daily and weekly hours of work for minors and other exceptions.
6. Waiting time as hours of work.
7. Rest and meal periods as hours of work.
8. Sleeping time as hours of work (for example, central station monitoring personnel or emergency reporting personnel).
9. Meetings, lectures and training programs compensation.
10. Travel time as hours of work.
11. Settling grievances.
12. Civic and charitable work.
13. Medical attention.
14. Methods of recording hours of work.
15. Definition of absenteeism.
16. Policy on reporting absenteeism.
17. Policy on excused and unexcused absences.
18. Obligations and procedures of reporting absences.
19. Disciplinary policy on absenteeism.
20. Personnel forms used in controlling absenteeism.

Wages and Salaries

1. Basic wage and salary policies.
2. How wages and salaries are determined.
3. Wage and salary differentials.
4. Overtime pay policies.
5. Methods of payment.
6. Time of payment.
7. Payroll deductions.
8. Personnel forms and other policies used in administrating pay programs.

Seniority and Promotions

1. Definition of seniority.
2. Exceptions to seniority.
3. How seniority is lost and accumulated.
4. How seniority works relative to layoffs, recall, promotions, transfers, overtime, vacations, etc.
5. Seniority by classification.
6. Job bidding procedure.
7. Defining types of promotions.
8. Performance review.
9. How promotion affects seniority and pay.
10. Definition of transfer.
11. How transfers affect seniority and pay.
12. Definition of temporary and permanent layoff.
13. Reasons for layoff.
14. Advance notice of layoff.
15. How seniority is affected by layoff.
16. Layoff procedure.
17. Recall procedure.

Employee Benefits

1. Length of vacations.
2. Eligibility and amount of vacation pay.
3. Scheduling vacation time.
4. Illness during vacation.
5. Extending, accumulating or splitting vacation.
6. Military reserve or National Guard duty obligations.
7. Effective lay-off and absences on vacations.
8. Effective discharge and resignation on paid vacations.
9. Part-time and temporary employee paid vacations.
10. Number of paid and unpaid holidays.
11. Arrangements for religious holidays.
12. Holidays falling on non-working days.
13. Arrangements for long holiday weekends.
14. Eligibility requirements for paid holidays.
15. Amount of holiday pay and computation procedure.
16. Pay for worked holidays.
17. Eligibility and procedure for personal leaves.
18. Length of personal leaves.
19. Effect of personal leaves on earning, seniority and benefits.
20. Reemployment privileges and obligations.
21. Eligibility and procedure for military leaves.
22. Eligibility and procedure for maternity leaves.
23. Eligibility and procedure for jury duty and civic duties.
24. Sick leave—eligibility requirements.

25. Length of sick leave.
26. Amount of sick leave pay.
27. Proof of illness.
28. Accumulation of sick leave credit.
29. Unused sick leave.
30. Effect of sick leave on seniority and benefits.
31. Reemployment obligations for employee.
32. Defining group life insurance.
33. Eligibility requirements for insurance.
34. Amount of coverage of insurance.
35. Contributions or other.
36. Coverage of life insurance after employee's retirement.
37. Employees eligible for coverage of accident and sickness insurance, type and amount of benefit provided—contribution requirement.
38. Hospitalization insurance, type, eligibility, etc.
39. Profit-sharing plans.
40. Cash plans.
41. Wage dividend plans.
42. Production sharing or cost saving plans.
43. Deferred profit-sharing plans.
44. Pension plans.
45. Bonuses.
46. Employee service recognition and privileges.

Safety, Security and Operations Rules

1. How rules are to be publicized.
2. Type of penalty given employees for rule violation.
3. Type of disciplinary action.
4. Rules regarding first aid.
5. Arrangements for medical services.
6. Safety and accident prevention programs.
7. First aid training programs.
8. Health and safety clothing.
9. Sanitary and health facilities.

Complaints and Grievances

1. Defining the term grievance.
2. Steps to be followed in the grievance procedure.
3. Policies requiring grievances to be written out.
4. Policies regarding grievance sessions.
5. Policies regarding how employees are paid during grievances.
6. Payment for time spent in processing grievance.
7. Policy defining how arbitrators are to be selected, used and paid.

Termination of Employment

1. Description of conduct which warrants discharge.
2. Policies providing employees to be suspended before discharge.
3. Policy stating that management has a right to determine cause.
4. Policy and procedure to be followed in issuing written warnings.
5. Policy statement of management's right to discharge part-time, temporary or probationary employees.
6. Policy of when termination of employment notice is to be given.
7. Type or form of termination notice.
8. Contents of termination notice.
9. Policy in providing discharged employees with severance pay.
10. Policy on discharges being subject to grievance procedure.
11. Policy statement requiring or not requiring of employee notice of intent to resign.
12. Policy on a penalty provided for failure to notify the employer of intent to resign.
13. Policy on issuing letters of reference for terminated employees.
14. Policy in providing employees who quit or resign with termination form describing reason for termination.

Conclusions

Of course, the work we do in preparation of the material is a reward in and of itself. But ultimately, the real rewards will be with your people, customers, and suppliers knowing who to deal with, under what circumstances to deal with them and what to expect and what not to expect. What it all means is that your company will have developed greater predictability, greater stability and, of course, overall greater ability to face the challenges of the '80s!

The points that Ron Davis made in outlining material for company manuals are valid. Every company, large or small, should have its policies and procedures established in written form. Instead of two separate manuals, as Davis suggests, one might be all that is needed, especially for smaller companies. As your business grows and you add people to your staff, you will probably want to have two. A sample organization manual can be found in Appendix D at the end of the book.

TECHNIQUES II

Marketing Your Goods and Services 7

To market your wares you need prospects, potential customers who have a need for your products and services and the ability to pay for them. Unless your prospects have both the need and the ability to pay, you will waste a lot of time. In this chapter we explore some methods of locating prospects and techniques to assure that they are good ones.

Companies advertise for a variety of reasons: to project company image, increase brand awareness, and introduce a new product or build sales for an existing product, to mention just a few. Most alarm companies advertise with one primary purpose in mind: to get leads that a salesperson can follow up and attempt to make sales.

Advertising is not the only method to generate leads, however. We explore that as well as other techniques that can result in sales.

ADVERTISING

Advertising talks to groups rather than to individuals. You can tell more people about what you sell faster and at lower cost by advertising than by any other means.

Selecting the appropriate advertising medium requires careful planning. The object is to find the one that will reach your particular market in an effective and economical manner. In some cases that might require the use of several media. All have specific capabilities.

If you have defined your market, you should be able to determine the types of persons that your message should reach. Once that is established, selecting the correct medium will be simpler.

Advertising is designed to induce on the part of the reader or listener a favorable reaction to a product or service. Its purpose is to attract prospects to your organization by convincing them of the superiority of a product or service. Turning prospects into customers will be the job of your sales staff. Except for direct mail-order selling, advertising is not expected to close sales.

Action resulting from advertising may be direct and immediate or indirect and delayed. The response of a person to supermarket advertising is direct action. The slower response of the person who buys an alarm system is indirect action.

Direct action can be expected for items purchased often in stores. Soon after the advertising appears, the results show up in sales. For products not purchased often or not urgently needed, buyers are likely to wait until a convenient time to act. Or they may be satisfied for the time being, but will remember your advertising if they decide later to purchase a system.

Advertising requires four major decisions: what to tell in your message, to whom to tell it, how to tell it, and where to tell it. Our discussion is devoted chiefly to helping you decide where to tell it, that is, in what media to advertise.

Advertising Budget

Spending too much money without getting results in extra sales can be costly. Spending too little without getting adequate penetration of your potential market area is wasteful. The business maxim "You've got to spend money to make money" is certainly true in advertising.

Total yearly expenditures for advertising are usually expressed as a percentage of gross sales. If you had $100,000 in gross sales last year and spent $2,500 for advertising, your advertising budget was 2.5% of sales. Depending on your particular operation, your expenditures probably should range between 1.5% and 3.5% of sales, though it is possible that a larger percentage might be needed.

The gross-sales method is not the only guide for establishing your advertising budget, although it is the one most commonly used and most easily understood. If your company is new, a larger budget may be required for you to become known throughout your market area. Also, you cannot ignore the amount of advertising done by your competitors. The more they do, the more you should do.

If you are using two or more media for advertising, it is a good idea to plan your monthly expenditures for each medium and to record the information on a calendar. The monthly budget for each medium, the size of the ad, its cost, and the day on which it will run should be shown on the calendar. The items or services to be promoted should be listed too.

Selecting an effective medium for advertising is not a simple job, though it is easier if you know your intended market. Table 7.1 gives general information on the coverage, advantages, and disadvantages of the various media.

Time and space costs are related to the following variables for newspaper and radio advertising:

1. Newspapers generally quote their rates for local display advertising by column inches. If line rates are quoted, remember that there are fourteen lines to the inch. Rates are almost always higher for national than for local advertisers; thus you, as a local dealer, should be entitled to some of the lowest rates offered. Also, most local newspaper rates are based on volume, which means that as the amount of space you contract for increases, the cost per inch or line decreases.

2. If you ask to have your ad on a certain page of the paper or in a specific position on the page, the paper may grant your request as a matter of goodwill. To guarantee a special position, however, you will probably have to pay a premium rate. The general practice is to charge premium rates for advertising on pages near the front of the paper, such as the second and third pages, and in special-interest sections, such as women's sections and sports pages. Premium rates also are charged for color.

3. Classified rates are quoted by lines. Occasional users pay transient rates, but a business user usually will qualify for professional rates, which are lower. These rates, whether for regular or display classified advertisements, are usually based on both frequency and volume. The frequency rate may be for consecutive days or for the total number of days, consecutive or otherwise.

4. Radio time is sold in ten-, thirty-, or sixty-second spots. A useful buy is the "adjacency," which gives small businesses a spot immediately preceding or following a popular feature (such as news, sports, or weather) but without the financial commitment of total sponsorship. If radio is your principal medium,

Table 7.1 Advertising Media Comparison

Medium	*Market Coverage*	*Type of Audience*	*Advantage*	*Disadvantage*
Daily newspaper	Single community or entire metropolitan area; sometimes zoned editions	General; tends to be older age group with slightly above-average income and education	Flexibility	Unable to select specific audience
Weekly newspaper	Usually a single community; sometimes metropolitan area	General; typically residents of smaller communities	Local paper identifies with local businesses	Limited market coverage
Shopper	Controlled by advertiser	Consumer households	Consumer oriented	A throw-away or give-away and not always read
Directories (including *Yellow Pages*)	Special occupational fields or specific geographic area	Active shoppers for goods and services	Users are in the market for goods or services	Limited to active shopper
Direct mail	Controlled by advertiser	Controlled by advertiser	Personalized approach to an audience of good prospects	Several variables can affect success
Radio	Broadcast area	General, unless station has a special programming format	Limited market selectivity: mass appeal	Should be used frequently to be effective
Television	Broadcast area	Varies with time of day and programming; tends toward younger age group	Wide coverage, dramatic impact	Cost of time; cost of ad production
Transit	Area served by transit system	Transit riders, shoppers, drivers, pedestrians	Repetition and length of exposure	Limited audience
Outdoor	Advertiser selectable	General, especially drivers	Dominant size, frequency of exposure	Cost

one-minute spots are usually preferable; if it supplements other media, shorter ones may be acceptable. The number of spots and when and where they are placed will be governed by how much you spend. As a general rule, the more spots the better. One of the strong points of radio advertising is the opportunity for frequent repetition.

5. Radio time is divided into several classes. The highest class, prime time, is the most expensive. Radio prime time corresponds to morning and evening drive times. As with other media, rate structures vary according to the size of the audience a station can reach. Except on the smallest stations, time periods are classified for the purpose of setting rates in accordance with the following plan:

Class AA	Morning drive time—6 A.M. to 10 A.M.
Class B	Homemaker time—10 A.M. to 4 P.M.
Class A	Evening drive time—4 P.M. to 7 P.M.
Class C	Evening—7 P.M. to midnight
Class D	Nighttime—midnight to 6 A.M.

(The hours shown are not exactly the same for all stations.)

Before you examine the value of any media, however, decide what you want to accomplish. Then take a look at what each one has to offer you as an alarm dealer.

NEWSPAPERS

Most advertising in daily newspapers is local. That is, the ads are placed by local merchants rather than national companies. In weekly newspapers and shoppers, the advertising is almost entirely local.

Reading habits are changing. Many people now depend on radio and television for national and international news, but some still read the big-city daily newspapers. They also want news closer to home and of more personal interest, however, and for this type of information, they rely on their local newspapers. Many big-city dailies are now issuing zone and suburban editions in response to these habits.

People of all ages, both sexes, and various educational and income levels read newspapers. Advertising therein is bound to reach some in your area who are logical customers.

The flexibility of newspaper advertising is a major advantage to that medium. You can let the paper insert your advertisement wherever there is a place for it (called ROP, or run of paper or run of press), or you can specify the page or section.

Your advertisement can be submitted only two or three days before the publication date, in most cases. This allows you to change your ad quickly, if necessary.

Dailies

If you live in a large metropolitan area you may have a choice of several daily papers for your advertisements unless you plan to advertise in all of them. You will need to decide which paper reaches more of the people you are interested in and how the advertising costs compare. If you live in a smaller community with only one paper, the selection has been made for you.

Nationwide, evening editions outnumber morning editions. They are either delivered to the reader's home or picked up on the way home. They are family oriented, read at leisure by both husband and wife as well as by other members of the family.

Morning papers tend to be read away from the home and are often bought at a newstand on the way to work. If home delivered, they may be taken to work or read on the bus or train.

Depending on whether you are marketing residential or commercial systems, the choice of an evening or morning paper could make a difference in the impact of your advertisement.

Sunday newspapers are of two kinds. One is another day's issue of a morning or evening daily but thicker, with more special sections. It provides a high degree of selectivity for reaching people with different business, home, and recreational interests. The other type is really a weekly. Both have the advantage of reaching the reader on a day when he has more time for reading than during the week.

Weeklies

Weekly papers include some that are published more than once a week, but they are classified as weeklies because they do not appear daily.

Rural weeklies are published in small communities at some distance from cities of any size. They are usually the only newspapers practicable for local advertising in those areas.

Surburban weeklies, however, are read in areas where dailies circulate. Depending on their circulation and audience, they should be considered as an additional medium or, possibly even as the primary newspaper for your advertising.

Shoppers

The weekly shoppers are similar to newspapers in some ways and different in others; they are sometimes called shopping guides or ad papers. Shoppers are usually delivered or distributed at no cost to the reader. Some contain only advertising, others contain a mix of ads and editorial material, with advertising being the predominant portion.

How well shoppers fit into your advertising plans will depend on the market you are trying to reach. Do not ignore them. Judge them on their merits as you would any other newspaper.

Quantity and Frequency

If you advertise in a weekly, it is probably best to run an ad every week, though every other week will still give you continuity. If you advertise in a daily, you have a wider choice.

Wednesday, Thursday, Friday, and Sunday are days of heavy advertising. Though you may fear that your ad will get lost among the others, especially if they are larger, those are the days when people plan to buy certain products. It is likely, though, that alarm systems are not among those products. The day your ad runs probably is not a critical factor, except that you should plan it so that it does not compete with too many other ads. Monday, Tuesday, or Saturday may be your best choice.

Size

The size of your advertisements will probably be related to their frequency and the funds you have available. There are other factors to consider, though.

How large a specific ad will be depends on how many products and services you mention, how large the illustrations are, how many words of description you use, and how bold your headline will be.

Classified Advertising

Do not overlook the classified pages, especially if your advertising dollars are in short supply. Some classified advertising can be used to supplement your other ads, or you may choose to use it exclusively if that is where you obtain the most response.

Classified sections take either regular or display ads. Even with restrictions as to type styles and sizes and space units, some very effective, attention-getting lay-outs can be made up with display classified advertisements.

Cost Comparison

To find out what your advertising is costing you per reader contact, multiply the circulation by the number of times you advertised during the year. Then divide the amount you spent for the year by the figure you obtained in the first step.

To compare the costs of advertising in different newspapers, it is customary to use the cost per line per million circulation, called the milline rate. Use the highest rate charged by each newspaper—not the rate you are contracting for—in the following formula:

$$\frac{\text{Line rate} \times 1 \text{ million}}{\text{Circulation}} = \text{milline rate.}$$

The reason for multiplying by 1 million is that larger figures are easier to compare. If the rates you are comparing are quoted in column inches, they can be used in the formula instead of the line rate. Just be sure to use the same rate base for all the newspapers you are comparing.

DIRECTORIES

In directory advertising, you reach people who have already decided what to buy; they want to know where to buy it. They are ready-made prospects. There are numerous types of directories in addition to the telephone book's *Yellow Pages.*

In its simplest form, directory advertising lists your name and address under alphabetical headings of products and services. Your listing may be free or you may be charged for it; sometimes just the listing of your name and address under a product or service classification is free. Publishers want their directories to be as useful as possible, and this requires a complete coverage of sources of supply.

Some directories provide for advertising as a part of your alphabetical listing. This type of display advertising is useful in two ways. First, near your listing you can tell more about yourself and what products and services you offer. Second, the advertisement, both in size and content, may have a psychological effect on the reader by letting him know that you are interested in his business. Your display ad may well be the reason he decides to call you.

Yellow Pages

According to a recent survey, 75% of all adults use the *Yellow Pages.* The study noted that half of them do not have a particular company in mind when they use the directory to locate a product or service.

The strength of your advertising message in the *Yellow Pages* will determine how many calls you will receive. Grab the user's attention with a well-conceived headline. Then tell him the important points about your business, those that explain why he should choose you over your competitors. Finally, make it easy for him to call or visit your company. Make your telephone number and address, with directions if necessary, easy to find in the ad and encourage him to call.

In the *Yellow Pages,* one listing is provided free of charge because you are a business-telephone subscriber. Additional listings cost extra. As a telephone subscriber you will also have a listing in the white pages. If your white pages listing is in boldface type, you may be wasting money if you are paying extra for the larger, heavier typeface. While the boldface looks good, it has no effect on sales. If a prospect is looking for your company by name in the white pages, he will find it even if it is in regular type. If he does not know your company name, he will not even be looking in the white pages.

DIRECT MAIL

The difference between direct-mail advertising and its relative, direct advertising, is the method of delivery. Direct-mail advertising is sent through the mail; direct advertising is handed out by salespeople or distributed house to house. One or the other or both are used by most alarm companies.

Direct-mail advertising is the most selective and flexible of all media, selective because you decide who is to receive it. You advertise only to people who can use what you sell. Your ads can be sent to a few people or to thousands. You can confine it to a small area, such as a few city blocks, or distribute it to your entire marketing area. It is flexible because the sizes and shapes of your ad are up to you; the presentation can be simple or elaborate. You distribute it when you choose; you are independent of publication dates and broadcast schedules. You have almost complete control over the medium.

Another advantage is that the reader is not distracted by other ads; while yours is in his hands, you have, at least for the moment, his undivided attention. Direct-mail advertising has to be good in what it offers and in how the offer is presented for the prospect to take time to read it, however.

Direct-mail advertisements can be made personal and specific because they are addressed to individual prospects rather than to heterogeneous groups. Letters, in particular, afford you an opportunity to talk candidly and directly to known prospects about merchandise in which they are interested.

This material stands an especially good chance of receiving favorable response because it can be sent to the home or wherever addressees will be most receptive, and it may be timed to reach people at opportune moments. Results can be traced more accurately than with any other medium because purchasers' names can be checked against mailing lists.

Uses of Direct-Mail Advertising

Direct-mail advertising has several uses for alarm dealers:

1. To presell prospects before a salesperson's call; to acquaint him with your company and its products and services
2. To announce new items or changes in your products and services
3. To follow up on a salesperson's call to prospects
4. To welcome new customers
5. To thank all customers for their business at least once a year

6. To remind customers of seasonal or periodic needs such as testing their alarm systems monthly
7. To take advantage of printed advertising materials supplied by equipment manufacturers

If you doubt that you are reaching all prospects through other advertising media, direct mail gives you another opportunity. If you are already reaching them, it adds impact to your message.

Types of Direct-Mail Advertising

Direct-mail advertising in its very simplest form is a postal card with a message handwritten on it. It also can be an elaborate self-mailer or an envelope with numerous enclosures.

Sales Letter. The sales letter is a simple and personal direct medium. Because it can be made to resemble a personal communication, the letter will at least be opened and given an opportunity to attract the attention of the addressee.

If only a few letters are to be sent out, they may be typewritten singly. Large numbers of letters, however, can be produced more rapidly and at lower cost by having them typed individually on an automatic typewriter or word processor. The personal effect may be enhanced by inserting addressees' names at appropriate places in the body of the letter.

Envelope Enclosure. An envelope enclosure is the least expensive form of direct-mail advertising because it can be included in letters, invoices, or statements. If the combined weight of the communication and the enclosure is kept under one ounce, allowing it to be mailed for the minimum first-class rate, the envelope enclosure makes your regular expenditure for business postage serve as an investment in advertising, too. It is useful in advertising new products and services to present customers, since it is sent out with your monthly billings.

While the term "envelope enclosure" may be applied to a variety of advertising material, it most often describes folded circulars and small unfolded stuffers.

The circular, a popular form of enclosure, is effective because it may be sufficiently large to enable you to tell all the essential facts about one or more of your products or services. If a good-quality paper is selected, you may employ any kind of illustrative or typographic display that is appropriate.

The small stuffer, or leaflet, also can prove valuable. Brief new product announcements may be made, or periodic security information provided as goodwill, such as a list of vacation security hints, works well.

Postal and Mailing Cards. The Postal Service's postal card is a satisfactory, inexpensive medium for sending brief advertising messages. The modest price includes both postage and card, yet it is treated as first-class mail. There may be an advantage to the postal card because it exposes its message to the recipient instantly. Should you desire a response, a double card is available, one-half of which is detachable and can be returned immediately since postage is already included.

While it may be a minor, technical point, it should be noted that a *postal* card is one provided by the United States Postal Service (USPS) and has postage already on it. A *post card* is one purchased from any other source and must have a stamp or other form of postage payment applied to it.

Mailing cards are post cards. You may select suitable grades of cardboard and, for a special effect, employ a variety of illustrative techniques. Private mailing cards must conform to the size and form regulations prescribed by the USPS to be admissable at the lower post card rate.

Business-reply Cards and Envelopes. The old practice of enclosing government postal cards in advertisements was a wasteful method of encouraging prospects and customers to respond because so few were returned. The advertiser had to pay postage on the unreturned as well as the returned cards. Business-reply cards are now commonplace. Neither you, the advertiser, nor the addressee is required to put a stamp on the card. Instead, you pay the post office a moderate fee in addition to the regular first-class postage for each card that is *returned.* A special business-reply permit is required and can be obtained from your local postmaster.

Business-reply envelopes have largely replaced stamped reply envelopes. Similar to business-reply cards, a moderate fee in addition to postage is charged on the returned envelopes. Some advertisers continue to use stamped reply envelopes, however, because they believe that addressees feel more obligated to respond when advertisers actually spend their money for return postage.

Self-Mailing Folder. The self-mailing folder resembles a circular or envelope-enclosed folder, but is usually made of heavier stock paper so that it can be folded and mailed without an envelope. It may be used to carry an advertising or goodwill message.

Broadside. The broadside is a giant-sized folder, designed to impress by its sheer physical size and striking display. As the broadside is opened, the successive surfaces present a continuous selling story, culminating in a huge single-sheet spread that hides anything else from sight. Sufficient space is available to tell a complete sales story. It is advisable to present only the essential facts about the product or service that you are promoting and to display these facts in large-sized type, to get the selling point across quickly.

Mailing List

The strongest reason for using direct mail is the degree to which you can select your audience. You can do it by sex, age, income, educational level, occupation, or almost any characteristic you wish. (An extreme illustration of the selectivity that can be achieved is the list of men six feet tall or more and weighing at least 215 pounds.)

The key to selectivity is the mailing list. The one you use must be made up of people who can use your goods and services and who have the ability to buy them or can influence others to purchase them.

To compile your own mailing list, consider these sources:

1. Present customers: you will want to inform them of new products and services or encourage them to upgrade their systems.

2. Names turned in by salespeople: these could be people who are going to be visited or have just been called on or they could be leads generated by your sales staff during a sales presentation.

3. Telephone directories: in a criss-cross directory, names are listed in a reverse order. That is, you select prospects by geographic area, street addresses. You can make up a list of just a few blocks or for a larger area. The *Yellow Pages* section is useful if you want names and addresses of specific types of businesses.

4. Civic, social, and professional directories: membership rosters are a good source of names and addresses.

5. Government records: the offices of the city and county clerks have a wealth of information if you are willing to spend some time searching for it. For example, building permit applications will give you an idea of what and when new construction is being planned.

Additional names for mailing lists can be found in newspapers and from noncompetitive businesses. You also can ask your present customers to suggest people who should be included.

Mailing List Companies. Many companies make a business of compiling mailing lists. These lists can be general or specific.

Compilers guarantee their lists to be current up to a specific percentage. They refund postage on pieces returned as undeliverable beyond the percentage allowed by the guarantee. For example, on a 95% guaranteed list, postage is refunded on returns above 5% of the list. A list purchased from one of these companies may be better than one you make up yourself and may cost less in the long run. It will certainly save you a lot of work.

Mailing List Maintenance. Even the best of mailing lists deteriorate, maybe by as much as 25% per year. The addressees get married, change their addresses, get promoted, change employers, or die. Companies go out of business, merge, or make other changes. So keeping the list up to date is important. To keep your list current:

- Be sure that all changes in your billing records are made in the mailing list.
- If your list is made from sources such as telephone or trade directories, check it every time a new issue becomes available.
- Have your sales staff report personnel changes in customer companies.

Periodically, send out mailings that include "address correction requested" under the return address. The post office will inform you of the new address or the reason that the piece was undeliverable. A small fee is collected for this information, but it will keep you from sending mail to persons who are no longer prospects.

Postal Rates. How will your direct-mail advertisements be sent? First class? Third class? First class is faster, but it costs more.

Third-class mailings may be made as single pieces or as bulk mail. Bulk mailings are cheaper, but a permit is necessary and certain regulations must be complied with. Typically, bulk rate mailings require a fee of about $40 per year and the mailings must be in batches of at least 200 pieces and sorted by ZIP code. Bulk rate mailings are less expensive. Individual pieces weighing several ounces can be mailed for slightly more than half the first-class rate for a one-ounce piece.

Pitfalls to Avoid. Three common mistakes are made by direct-mail advertisers.

1. Mailing on Friday: Friday is the worst day to send out mailings to businesses and sometimes the worst for home-addressed mailings. Higher returns from business-addressed mail result when the mail is received on Tuesday, Wednesday, or Thursday. A businessperson has much more mail to get through on Monday, and may devote less time to something new. On Fridays, he may be eager to finish up the week and be less likely to be receptive to a new idea. Home mailings received on Friday or Saturday are most productive because the homeowner has plenty of time to read your advertisement. A letter mailed on Friday *may* arrive on Saturday, but why take the chance that it will not? Mail to residential prospects on Thursday.
2. Impersonal material: the closer the mailing resembles a personal letter, the more responses you will get. For example, a letter addressed to a specific person will be better received than one addressed "Dear Friend." Moreover, a letter without an enclosure may be effective. An enclosure without a letter probably will not be effective.
3. Not making it easy to respond: responses can be increased by enclosing a reply card and even more so if the card or enclosed envelope is already self-addressed and stamped. Better still is the reply card on which the addressee's name is already printed, so all he needs to do is drop it in a mail box.

RADIO

Radio advertising is almost entirely local advertising. As a medium for locally oriented sales and service providers, its value is probably surpassed only by that of newspapers.

The human voice over the air can establish a friendly rapport with listeners. It can be more persuasive than a printed message; it also can convey urgency. If you want an immediate response, you can end your advertising message with a suggestion that listeners call right away. Spoken in a conversational manner and repeated frequently, radio advertising is remembered.

By selecting the station, program, and time of broadcast, you can reach almost any group of buyers that you want. In choosing a station, keep in mind that your selection should not be based on what you like to listen to, but what your prospective buyers listen to.

Sponsorship or Spots?

You can choose among sponsoring your own program, sponsoring a station feature, and using spot announcements in a variety of ways.

Sponsored programs have the advantages of prestige and individual identification. Besides syndicated programs, stations often have local-talent programs that can be sponsored; local talent usually means strong local appeal.

Sponsored features that are usually available are newscasts, weather and traffic reports, sports reviews, and similar features. You might sponsor a feature daily or on alternate days.

A radio spot is a commercial announcement lasting one minute or less. It offers more flexibility than sponsored programs or features. The number of spots and when and where they are placed will be determined, to a large degree, by how much you are willing to spend. Classification of radio broadcast hours was given earlier in this chapter. Rates vary with the time and popularity of the program or feature.

TELEVISION

Television as an advertising medium is used primarily by national advertisers and a few large local alarm dealers. Eventually, it will probably be used by more small dealers.

Television advertising may not be as expensive as you think. Network programs may be too costly for your advertising budget, but think in terms of smaller stations with smaller market areas, of spot announcements, and of time periods other than prime time.

Advertising on television is the closest of all advertising to personal selling. The combination of sight, sound, and motion comes into the prospect's home to deliver your message. You can explain and demonstrate your products and services to thousands of people simultaneously.

Planning television advertising is similar to planning radio advertising. You still want to select a station that reaches the most likely prospects for your company.

Prime time is usually between 7:00 P.M. and 11:00 P.M. (EST). Time is sold for sponsored programs and for spot announcements. Spot announcements are usually twenty, thirty, or forty seconds or one minute, though, occasionally, a ten-second spot may be purchased.

MISCELLANEOUS MEDIA

Transit

Transit advertising involves a variety of forms. Usually, it is available only in larger metropolitan areas.

Car cards of standard sizes in buses, trains, or other public transportation vehicles are sold as runs. A full run is a card in every vehicle in the system; a half-run is a card in half of the vehicles. Sometimes, specific routes may be selected.

The larger displays outside taxis and buses are traveling displays, sometimes called moving billboards. Rates vary with the size and position of the display.

Outdoor

According to surveys, outdoor advertising has high-frequency exposure—almost thirty times a month—because much traveling is done over the same routes. Whether it is a billboard along a highway or the side of your building, be sure that your outdoor advertising conforms to government requirements.

Specialty Items

Calendars. Calendars are probably the best specialty advertising item. They remind prospects and customers daily of your company and the products and services that you offer. The design is very important. The calendar must be attractive and still display your message if it is to be hung on someone's wall or placed on a desk.

Book Matches. Another popular item is book matches. They may not be effective if they are passed out indiscriminately, though. Plan where they will be distributed, in places where your likely prospects will be.

Pens and Pencils. These items may be too popular. Whether or not they are effective depends on who receives them. Your advertising message is lost if they are left in a desk drawer with dozens of others.

ADS THAT SELL

For an advertisement to be effective, a key element must be present: "a promise of benefit." When people look at any kind of advertising or promotion, there is only one question in their minds: "What's in it for me?"

Instead of emphasizing your company name or a product in the headline, emphasize what the product will do for the customer. What you want to do is sell the benefits of your products and services—not the products and services themselves.

For example, a headline that reads "Peace of mind can be yours" will probably get more attention than saying "Residential alarm systems from XYZ Alarms are custom designed." The former tells the reader that he will receive a benefit; the latter tells him who you are, which may be of little interest to him at the moment.

Let's face it, we are all here to sell products and services. Few of us are giant companies with dollars to spare on image building and nonselling messages. Save that for later.

Creating the Ad

The items that follow are important components for every ad that you prepare for the print media. The reason we include most of this information is obvious: to give the reader of your ad every bit of information needed to visit your showroom or call for an appointment.

Border. Find a border that attracts attention without detracting from what is inside it. It should stand out and be in good taste.

Headline. The headline should be a simple, brief statement that contains a promise of benefit.

Subheadline. The subhead reinforces the benefit and adds new information.

Copy. Having attracted the reader's attention with the headline and stirred his interest with the subheadline, the ad copy—the written portion that provides amplification and details about the promise of benefit—should give the reader information about your products and services. It should be enough to amplify the promise of benefit and pique his interest so that he will call you for more information.

It is important to remember that you are attempting to sell the benefits of your products and services, not the products and services themselves. You are selling security, safety, and peace of mind, not a control panel and other hardware.

The reader might question whether or not you are an unbiased source of information. As an advertiser, you are expected to tout your own products and services over those of other dealers. To add credibility to your advertising message, you should include facts that will serve as proof and will reassure the reader that you are capable of fulfilling the promise you have made.

Some type of evidence, such as your length of time in business, number of satisfied customers, professional alarm association memberships, or Underwriters Laboratories (UL) listing, is worth mentioning.

Photograph or Illustration. The photograph or illustration should be attention-getting. When your ad is on a newspaper page with other advertisements, a good way to gain attention is with a well-placed illustration.

Price. Including prices in alarm company ads is a controversial subject. Some dealers advertise prices and some do not.

It is difficult to include a specific price for an alarm system because of the variables: equipment, labor, service contract, lease fees, and so forth. An argument in favor of listing a starting price is to qualify the people who call in response to the ad. For example, say your ad is to obtain leads for a residential alarm system and your typical system sells for about $1,500. You may get calls from people who can only afford $500. You have to answer every call and spend time explaining all the details to everyone. By not putting the price in the ad, you have wasted a lot of time.

The way you state the price may overcome this problem. Instead of listing a price of $1,500 for a system, you can try $35 per month on a lease basis, making it more attractive to more people.

Company Name or Logo. Your name and logo should be distinctive and easy to recognize. After you have run several ads, people should be able to identify your ad just by the logo. You want to build recognition.

Slogan. Similar to your logo, your company slogan should be easily recognizable, something short and sweet that people will remember. A word of caution: avoid the word *best*—it has been overused and tends to be counterproductive. One alarm company's slogan, "Your security is our business," is effective and simple.

Address. If you have a showroom and are attempting to entice people to visit, you may need a few more details than just a simple address. If necessary, give a landmark (inside the Suburban Mall), directions (one block east of Main Street on Oak Avenue) or include a map if necessary.

Telephone. Like the address, your telephone number is critical. People cannot call to set up an appointment if they have trouble locating your telephone number.

Charge Cards. If you accept charge cards, mention that fact. It is another way of making your products and services affordable.

Related Products and Services. If you do more than sell and install alarm systems, note the related products and services. For example, if your company provides locking hardware, you might say "twenty-four-hour locksmith service" or something similar.

Ad Layout. In planning the layout, look through several copies of the newspaper if that is where your ad will be placed. Advertising is somewhat imitative. Good styles, arrangements, and formats can be adapted to suit your needs.

To design your own ad, you will need thick and thin markers, a pencil, plain white paper, and tracing paper.

Start by framing out a size that will fit your budget. Draw the border on the white paper. About one-fourth of the way up from the bottom, on the white paper, sketch your logo, slogan, and other company information.

Tape a sheet of tracing paper over the white paper that you have just prepared. Keeping the white sheet as your "master" saves you the time of redoing the border and logo every time you want to sketch another ad. At this point, about three-fourths of the page is blank.

Sketch a big, bold headline with the thick marker near the top of the page (tracing paper). You may want to sketch the headline with a pencil, and then when the placement is satisfactory, use the marker.

Next, draw a block for the illustration, or if you have the illustration handy, make a copy of it or trace it. In the space that is left, insert a subheadline and pencil in the copy (the detailed message of your ad).

Because you have used tracing paper, you can remove the first version and do another. Prepare several. Then select the one that best projects your message. When you are finished, you are ready to call the newspaper or printer for the rest of the work.

Printers and newspapers have a variety of type styles from which to choose. Ask to see samples. Stick with one type style for the entire ad, and vary the size and boldness of the type. If possible, use a sans serif (without serifs) type style like Helvetica, Helios, or Avant Garde, which shows up well and is easy to read.

Do It Yourself

It is possible to do the entire ad layout by yourself. If you have crisp, clean black-and-white illustrations and a lot of patience, you can design a professional looking advertisement. Dry transfer, commonly known as rub-on, letters are available at most office supply stores. They come in a variety of type styles and sizes and, with some practice, are simple to use.

Some manufacturers have black-and-white illustrations of their products. If you use a specific product regularly, contact the manufacturer about the availability of illustrations.

COOPERATIVE ADVERTISING

The exact figures for cooperative advertising expenditures are impossible to calculate because much of the information is confidential. It is estimated that $1 of every $6 spent for advertising is cooperative, however.

Sharing the cost of advertising, that is, the product's manufacturer pays a portion of the advertisements of a local dealer, is commonplace in some industries. Unfortunately, the alarm industry is not among them.

A manufacturer advertises to tell people what to buy. If he shares the advertising costs with a local dealer, he can then tell them where to buy his product. Moreover, since local advertising rates are less than national rates, the manufacturer gets more for his advertising dollar. For the dealer, it is an incentive to stock and advertise the manufacturer's goods.

If you are selling large quantities of products made by one or two manufacturers, it would be advantageous to both of you to share advertising costs. Ask them about cooperative advertising. The worst that can happen is that they will say no.

ADDITIONAL INFORMATION

Only a few of the many sources of information about advertising can be mentioned here. Check with your local library for additional sources.

Media Directories

Standard Rate and Data Services, Inc., 5201 Old Orchard Road, Skokie, IL 60076, supplies basic information for approximately 14,000 media in a variety of directories, many of which are available in libraries:

Business Publications Rates and Data
Consumer Magazine and Farm Publication Rates and Data
Direct Mail Lists Rates and Data
Network Rates and Data
Newspaper Circulation Analyses
Newspaper Rates and Data
Spot Radio Rates and Data
Spot Television Rates and Data
Transit Advertising Rates and Data
Weekly Newspaper Rates and Data

Associations

Although specific associations emphasize the benefits of the media they represent, they are still good sources of information.

American Newspaper Publishers Association, Bureau of Advertising, 750 Third Avenue, New York, NY 10017
Direct Mail Advertising Association, 230 Park Avenue, New York, NY 10017
Institute of Outdoor Advertising, 625 Madison Avenue, New York, NY 10022
Radio Advertising Bureau, 116 East 55th Street, New York, NY 10022
Television Bureau of Advertising, 1 Rockefeller Center, New York, NY 10022
Transit Advertising Association, 500 Fifth Avenue, New York, NY 10036

United States Postal Service

Several publications are available from the USPS (check with your local postmaster):

Domestic Postage Rates and Fees
How to Address Mail
How to Prepare Second- and Third-Class Mailings
Mailing Permits

Government Agencies

The Federal Trade Commission and Small Business Administration offer some helpful material relating to advertising. Contact them and ask for a list of the publications that are available.

Federal Trade Commission, Division of Legal and Public Records, Washington, DC 20580

Small Business Administration, Washington, DC 20416

OTHER SOURCES OF PROSPECTS

In addition to advertising to prospects, leads can be obtained in other ways. Whether or not you decide to use any of the following sources will depend on your particular operation.

Referrals

Referrals by present customers are excellent leads and should be followed up quickly. Your customer is serving as a center of influence by recommending you, which means that you have a better chance of making a sale than if the lead were generated from other marketing efforts.

Newspaper Stories

Local newspapers, in the course of reporting the local news, will report robberies, burglaries, and fires. Often they include the names and addresses of the people in the stories. Using a criss-cross directory will provide you with the names and addresses of neighbors. People are much more likely to consider the purchase of an alarm system if a crime has occurred against someone they know.

Police and Fire Radio Broadcasts

The use of a scanner to monitor local police and fire broadcasts has a legal and moral aspect. First, some communities prohibit or restrict the use of these radios. Second, contacting a person within minutes of a burglary call being broadcast is an unprofessional action. At least give him an opportunity to complete the police report and recover from the shock of having his home or business invaded.

Making a Sales Call 8

As we discussed in Chapter 7, a prospect is a person or business who can both benefit from buying a product or service and can afford to buy it. A lead is the name of a person or organization who might possibly be a prospect. Some of the leads you develop will not become prospects, others will; still others will merely appear to be prospects. Those that appear to be prospects but, for whatever reason, are not are called "china eggs."

A china egg is one who appears to be a good prospect but who fails to "hatch" or turn into a buyer. The phrase is familiar to salespeople. It refers to the practice of farmers placing china eggs in the nests of hens that are not laying as a suggestion that they deposit their own eggs there, next to the china one. The hens often follow the suggestion, but meet with frustration when they try to hatch the fake egg along with the others. Salespersons often meet with a similar frustration when trying to hatch their china egg prospects.

The main reason the smart salesperson pays so much attention to prospecting is that it enables him to sell more. He can spend only a fraction of his time each day with prospects, so he is foolish to waste precious minutes talking with a nonprospect. One way to determine if the china egg will hatch is to make a sales call and present your products and services. If he buys, he has become a customer; if not, he should be placed with the rest of the leads that did not turn into prospects. There may be time later for a follow-up call, but it should not be made a high priority.

PREAPPROACH

Much work must be done before actually knocking on the prospect's door. The preapproach will provide you with information that will be useful for making a sales call.

There are five objectives to the preapproach: (1) to provide additional qualifying information, (2) to develop an approach strategy, (3) to obtain information around which the presentation can be better planned, (4) to avoid serious errors, and (5) to develop self-confidence.

Qualifying Information

Although you may believe that a certain person or business is a prospect, closer examination may reveal that he is not. He may already have an alarm system or he may be financially unable to buy one from you. What you are attempting to determine with your preapproach is whether or not the lead really is a good one.

Approach Strategy

Not all prospects should be approached in the same manner. Some are easy to meet, others are quite difficult to contact. Some like a direct, businesslike approach, while others prefer an approach that is less formal. Timing may be important. A sound preapproach should reveal the best approach strategy.

Planning Information

A sales presentation can take many forms. It can be built around cost saving, unique product features, or special services. If the prospect is most interested in low cost, it would be a mistake to dwell on top-of-the-line protection. Similarly, if he is interested in high security and special features, appeals to economy may be useless. A good preapproach should furnish the salesperson some insights into which motives are most likely to move the prospect to action and what appeals could prove most effective.

A good sales presentation should be tailor-made for the prospect. That does not mean that a standard, or "canned," presentation should not be used. Indeed, a well-prepared and rehearsed presentation has real impact. Knowing what will be important to the prospect allows you to adjust a presentation to meet special needs.

Avoiding Errors

People have many idiosyncracies that must be humored if a presentation is to be made successfully. For example, some people dislike overly aggressive individuals and react unfavorably to any salesperson who tries to overpower them. Others react negatively to a salesperson who smokes during a presentation. Know your prospect's preferences.

Self-confidence

The unknown breeds fear and uncertainty. The salesperson who walks into the presentation unprepared or unaware of the prospect's personality and situation is likely to become uncertain, hesitant, and fearful of making mistakes. The one who has done a complete preapproach is confident that what he will say is right and that he has the key to making a sale.

How can a prospect buy from you if he has no confidence in you, and how can he have confidence in you if you show no confidence in yourself? Think about it.

Preapproach Information

As much information as possible should be gathered about your prospect. Most of the following items should be checked whether the prospect is buying for himself or for his business.

Name. Learn to spell it and pronounce it correctly. People are sensitive on this point, and a mistake can be costly.

Age. Older people respond to the respect they feel is due them. Younger people in top-level jobs appreciate recognition of the fact that they have achieved important and responsible positions.

Education. This may provide a topic of conversation. A college graduate usually likes to have this fact recognized. The self-made person is proud to have won his position without so much formal education.

Residence. This may reveal something of his social position and his friends.

Need. If he needs an alarm system, you should learn how he can best use it. If he does not need it, he is not a good prospect.

Ability to Pay. It will not be worth while for you to try to sell someone an expensive system if he cannot afford it.

Authority to Buy. Will the spouse or business partner have to be consulted? You will waste your time giving your presentation to someone who cannot authorize the purchase.

Family. Many residential alarm systems are sold because of the protection it affords to the prospect's family.

Best Time to Call. Almost every person has a routine and does not like to have it disturbed. Schedule your call when he will be most receptive to talking with you.

Occupation. When you call on a prospect at home, it helps to know what he does for a living. Is he self-employed? What type of business? For whom does he work? What is his position? Answering these questions will not only be helpful for your sales call, they may provide additional leads.

Recreation, Interests, and Hobbies. Knowing this type of information is helpful in two ways. First, it may help you get past the "cold spot"—the first couple of minutes in the presentation. Second, if he spends every weekend fishing or collects antiques, that information is important when you explain how your system will give him peace of mind while he is out in his boat or at an auction. Do not go overboard on the hobby topic, however. If you lose control of the interview, you will be wasting your time.

APPROACH

It's time to meet the prospect face to face. What happens from this point on depends largely on the approach. Studies have shown that sales are won or lost during the first few minutes of the interview. A good approach should gain the prospect's attention and awaken his interest in your proposition. Moreover, it should lead easily and smoothly into the sales presentation.

You should regard the task of gaining an interview in exactly the same light as you do making a sale. It is your task to sell an idea to the prospect, or to a subordinate who blocks your way to the prospect. The idea that you are selling is that you have information about products and services that will be beneficial.

Making Appointments

Professional salespeople have a difference of opinion regarding the advisability of making a definite appointment. Some think it may be a disadvantage to commit themselves to a definite time, such as 10:30, for if they are ten minutes late they might lose the interview entirely or run the risk of having the prospect think that they did not appreciate his time. If you are engaged in closing another sale at 10:25, you cannot stop to go to your next appointment without losing the sale. Sometimes, a bit of leeway is desirable. This may be obtained by making appointments elastic, that is, schedule them for "between 10:30 and 11:00" if that is acceptable with the prospects.

It may be difficult to make an elastic appointment with a busy executive whose day is filled with other commitments, so a specific time will be needed. You might start each morning and afternoon with a definite appointment with a high-quality prospect. Not knowing how long the call will take, you would not make any other commitments for the day, but fill in any time left with cold calls on likely prospects.

Success follows the salesperson who possesses courage, courtesy, and confidence. The aspect of courage needed here is that which brings him back to the same prospect after repeated rejections. There is a story about a salesperson who was asked how many times he called on a prospect before crossing him off his list. "That depends on which of us dies first," the dealer said. Persistence pays off.

Courtesy shows itself chiefly in the interest that you have in the welfare of the prospect. If you know what your products and services can do for him, you will continue trying to sell him a system.

The confidence that helps get the salesperson past the numerous barriers must be rooted deep in his belief that he can do his prospect a real service. If you have that kind of confidence in yourself and in your products and services, you will convey that feeling.

Before making the approach to an important prospect, many salespeople indulge in a warm-up session much as athletes do. They review what they are going to say and do; they build up their enthusiasm; they don't enter the race cold, they are thoroughly steamed up.

SALES PRESENTATION

The presentation is the heart of selling. Everything to this point has been in preparation for it.

It may simplify the sales presentation if you think of it as a dual task: to arouse a feeling of need in the prospect's mind and then to show how your products and services fill that need. You may recognize that a need exists long before the prospect does, so be careful that you do not lose him during the presentation.

A good presentation must have four basic characteristics: (1) it should be complete, covering every needed fact; (2) it should eliminate competition, establishing your proposition as the best way to solve the prospect's problems; (3) it should be clear, leaving no misunderstanding in the prospect's mind; and (4) it should win the prospect's confidence that your statements are true and that you are thinking of his interests.

The memorized, standardized presentation is used by many alarm salespeople because it enables them to make the demonstration clearer and more complete, to win the buyer's confidence, and to eliminate competition more effectively. Other salespeople dislike the standardized talk and rely instead on the inspiration of the moment to guide their presentations.

Advocates of the prepared sales talk say that the salesperson is able to make his delivery with more spontaneity and enthusiasm if he is free to devote his attention to the method rather than to a search for words. Additionally, and, I believe, validly, there are several points that make a standardized presentation more effective:

1. It covers all the ground, leaving no gaps. It does not leave you wishing that you had remembered to tell the prospect about the standby power supply or the automatic bell cut-off. It is complete.

2. It ensures a logical order in the sales talk. The prospect is led from one point to another easily, the whole presentation building itself up into an effective pattern.

3. It saves time for you and the prospect.

4. It enables the beginner to sell effectively, whereas he might fail if left to his own devices.

5. It gives you more confidence since you know that you are prepared for a complete presentation.

Which method you select is a personal choice; however, it is strongly recommended that a prepared sales talk be used, at least until you have made some sales and have a steady income. After that, you can experiment as much as you want.

Preparing Your Own Standardized Presentation

If you want to prepare your own standardized presentation, you should do two things: (1) you should write it out in complete detail and then edit it carefully, and (2) when you think it is in its final form, you should speak it into a tape recorder and play it back over and over. You will probably make some revisions, after which you can make a new script, record it again, and listen to it again. Continue to make revisions until you are satisfied that the presentation covers all the ground in a logical order. The presentation is not a forum for a speech. It should be delivered as though it is spontaneous and extemporaneous.

It is especially important that major points, power phrases, and various closings be made verbatim. Another familiar sales anecdote is: "If you're playing the accordion just for fun, you can play by ear in your own way; but if you're playing for money, you'd better learn to play by note."

Organized Sales Talk

If you feel that the fully standardized presentation is too confining and inflexible, the organized sales talk may better serve your purpose. The difference is one of degree, because both follow a fixed pattern or outline; the organized talk allows more flexibility of wording and greater opportunity to linger on some point that appears to be making a good impression.

Prepare your organized talk in a manner similar to that of the standardized one. Write it out in complete detail and edit it carefully. Now, take the main topics that you want to cover and arrange them in a logical order, being certain that no important points are omitted. Review and revise the list of topics until it is complete. Repeat them until you are confident that you can proceed through the entire list without leaving out any points.

Clarity

Clarity means not merely clear enough to be understood, it means so crystal clear that it cannot be misunderstood. Here are four suggestions for making your presentation clear: (1) use showmanship, (2) gauge the prospect's comprehension, (3) use effective figures of speech, and (4) talk the prospect's language.

There are literally hundreds of ways to use showmanship in a sales presentation, but only a few are mentioned here. It is often advisable to use the dramatic to convey a message. It is not sufficient to tell, <u>show.</u> The eye is much more effective than the ear in conveying impressions to the brain.

Visualizing. The old style of selling without any aids is virtually obsolete. Good salespeople are now equipped with tools designed to tell the story quickly through the eye.

Compare the difference between watching a baseball game on television and listening to one on the radio. The radio sportscaster knows that his listeners cannot see anything happening in the park, while the television viewers can see a good deal and do not need or want the announcer to keep up a constant flow of chatter. When a salesperson uses any method of visualization, he does less talking.

Pictures of the control panel or certain sensors can be shown because the actual items would be too bulky to carry around. Smaller items, such as contact switches, can be carried in your briefcase and shown to the prospect.

Films, filmstrips, and slides provide another method for visualizing products. While they are effective, they require a projector and can be somewhat difficult to handle.

You may want to develop a sales portfolio that includes a collection of illustrations, graphs, letters from satisfied customers, records of tests, and other materials you believe will help to make an effective presentation.

The intangible service that you will be providing—security and peace of mind—needs effective visualization even more than tangible products. As a salesperson of an intangible, the use of a pencil and pad during the presentation may be effective. Many times, a simple chart or graph that is completed as the interview progresses adds clarity and maintains interest. The casual jotting down of a figure or a fact upon a pad also uses eye appeal successfully.

Dramatizing. Actually using some alarm components adds to sales talk by dramatizing certain points that you want to make. For example, a control panel's functions may be tested or a space protection device can be demonstrated. If you are using a demonstration kit or separate components, staging tests is relatively simple.

These tests should be simple. While it is a good idea to have the prospect participate in the dramatization, be sure that the chances for errors are minimized.

Planning

Building your prospect's confidence in you as well as in your products and services takes practice. You will make mistakes and some of them will lose sales. Learn from them. Always evaluate your presentations, looking for what you did that was right as well as what you did wrong. It is easy to fall into a rut when you use the same sales talk time after time. Plan your presentation carefully and keep it lively.

People will be interested in different aspects of your talk. Watch for clues that show the prospect is interested. Dwell on those aspects until he has been given enough information to answer his questions, then proceed to other topics.

Not all features of your system should be demonstrated. Stress those that are most favorable and do not dwell on ones possessed by all competitors. Moreover, it is easy for a prospect to become bored by a demonstration of minute details. It may be sufficient to tell him that the system has an entry/exit delay without explaining how many transistors are required to accomplish it.

Plan the pace of your presentation very carefully. Do not rush it. Because you thoroughly understand the system, do not assume that others do. When the concept is new to the prospect's mind, he must be led slowly until he has assimilated each aspect of it. If he is familiar with alarm systems, however, you may be able to proceed at a faster pace. Still, don't rush.

Reaction

Gauging your prospect's reactions takes time, just like building his confidence. To find out how clearly you have been putting across your selling points, it will probably be helpful to ask questions of your prospect as you go through the presentation. Questions like, "What do you think about that feature?" or "If you owned this system would you find this feature useful?" allow the prospect to tell you if he has understood what you have told him.

Avoid directly asking him "Do you understand?" It implies that he does not have the ability to comprehend what you have told him. Rather, ask, "Have I made myself clear?" That shifts the responsibility to you and away from your prospect.

Asking these questions serves more than one purpose. First, it reveals which features or arguments are most effective. If a feature is of interest to the prospect you can discuss it further; if not, move on to the next point. Second, by getting affirmative responses as you move through your presentation, you are blocking the prospect from raising objections to those points later.

Prospect's Language

Many salespeople do not talk their prospects' language. The importance of using short, simple words that convey your meaning cannot be overemphasized. You do not have to engage in a sesquipedalian diatribe to present your wares. (Sesquipedalian means "characterized by the use of long words" and diatribe means "a prolonged speech.") In other words, keep it simple.

The Lord's Prayer has 56 words; Lincoln spoke only 268 words when he delivered the Gettysburg Address; and our country was started on 1,322 words in the Declaration of Independence. Of the 268 words that Lincoln spoke, 196 of them had one syllable, 52 had two syllables, and only 20 had three or more syllables. (By contrast, recently it took the government 26,911 words to issue a regulation on the sale of cabbages.)

This does not mean to imply that for some prospects words of more than two syllables are ineffective. It does suggest a careful selection of language appropriate to each prospect.

Security Survey

At some point during the sales presentation you will need to survey the premises, that is, you will need to look at the building to be protected so you will know what type of alarm system to recommend. The method you use will depend on your company's operation. The important thing to remember is to be sure that you have not overlooked anything.

COMPETITION

Few sales are made without encountering competition of some sort. To render the competition harmless, you must be prepared. Do not be the one to bring up your competitors—you could be telling the prospect something that he does not know. If the prospect is unaware of other dealers and you inform him of them, he might call them for an estimate. If you will forgive a rather trite expression, let sleeping dogs lie.

It is your job to determine the competitive setting early in the presentation. Then choose your strategy for handling it. Generally, there are three good methods of doing this if the competition should be mentioned by the prospect.

Praise and Pass On

Praise the competition and pass on. If you ignore the other dealers entirely, the prospect is not led to consider them. Sell your own products and services, let the competition sell theirs.

There are a couple of drawbacks to this strategy. First, if a competitive brand or company is in the prospect's mind, ignoring it may do little to dislodge it. Second, the prospect may keep silent, knowing that you will point out the folly of his thinking if he mentions the other dealer he is considering.

Meet It Head On

There may be times when there is no graceful way to ignore the competition. When it becomes necessary to make a comparison between your products and services and another dealer's, try to avoid too much detail and do not attempt to cover every point, just discuss the ones that seem to interest the prospect. This, of course, involves knowing where you are strong and your competitor is weak, so do your homework before you attempt a comparison of this type.

Remember, too, that little is to be gained by speaking ill of the competition. Dwell on the strong points of your proposal, not on the weak points of the other fellow.

Recognize, but Handle with Care

The third view lies somewhere between the two other methods. It is wise to avoid vigorously attacking your competitors, and yet it may seem impossible to ignore them completely. One result of an attack is to instill in the prospect's mind that you are losing business to the other dealer, who therefore probably has a good proposition. You may have just lost the sale.

The best procedure is to acknowledge that competition exists, answer the prospect's question honestly and briefly, and move on to the system you are offering. Reinforce its positive aspects in which the prospect has displayed an interest.

NOW WHAT?

Some salespeople use what is called a one-call close. That is, they make their sales presentation, perform a security survey, and prepare a proposal for the prospect during their first call. If that method works for you, use it. The few times that I tried it, I was not pleased with the results. Hence, I use a two-call close. It is composed of two visits to the prospect. The first is to make the sales presentation and do the security survey. The second is to present a proposal and close the sale.

To be effective and to do a good job of designing a system for the prospect, I need time to think, and I cannot do that sitting in a prospect's living room or office. Back in my own office, I can unhurriedly design several systems and select the two best ones. Then I make another appointment with the prospect and prepare to close the sale on the second call.

This book is written based on the two-call close. This chapter has just covered the first call. Chapters 9 and 10 discuss designing the system and bidding the job. Chapter 11 addresses that all-important aspect of selling: closing the sale.

Designing the System 9

An alarm system must be designed carefully if it is to be security and cost effective. The design process need not be complicated, however, provided you have basic knowledge of equipment applications and installation techniques.

Before you decide what equipment you will use, you need to know where it will be used. Although components are discussed in detail in later chapters, a brief look at the three major components of an alarm system will help you understand the importance of careful planning (i.e., designing).

The three basic components are detectors, processing or control units, and annunciators. As the name implies, a detector detects something, such as a door opening, glass breaking, vibrations, pressure, movement, sound, and temperature, to name a few. For proper operation of the system, correct application of detection devices is imperative. False alarms that require repeated service calls can quickly erode the profit made on a sale. The rationale for selecting certain devices for specific applications will become apparent when the individual detectors are discussed in Chapter 14.

The processing unit, commonly referred to as a control panel, is the electronic or electromechanical brain of the system. Many new control panels, indicative of technologic advances in the industry, are microprocessor based. Applying computer technology to alarm equipment will provide systems that are more reliable and easier to install and service. The function of the processing unit is to receive a signal from a detector and respond. Selection of the proper control panel will depend on the number and types of detectors used, and the functions required of the system by the customer.

The processed alarm signal from the control panel is sent to an annunciator. The signal can be announced visually or audibly, or both. Bells and sirens are the most common forms of annunciation, although strobe lights are gaining popularity. Annunciation may take the form of a prerecorded message played on a telephone tape dialer or an electronic signal sent to a monitoring center.

SECURITY SURVEY

The first step in designing an alarm system is to survey the building to be protected. Walk through and around the building. Looking inside the premises will tell you the locations of doors and windows and provide you with basic information. Walking around the building may reveal some potential points of entry that are not readily apparent from inside. Occasionally, it may be necessary to check the roof physically for skylights and vents.

Diagrams or sketches of the building and/or individual rooms will prove helpful when you start designing the system. Though the diagrams need not be elaborate, the more detailed they are, the more useful they will be.

The following are typical questions that should be asked during a security survey. It is not necessary to ask the prospect or customer each one; many can be answered by simple observation. Some questions apply to residential alarms, some to commercial alarms, and some apply to both.

1. Doors: how many? exterior? interior? type? construction? location?
2. Windows: how many? type? construction? location? movable? nonmovable?
3. Walls: construction? exterior? interior? solid? hollow?
4. Floor/ceiling/roof: crawl spaces? suspended ceiling? Type of roof: hip? gable? vents? Could the building be entered from: under the floor? above the ceiling? through the roof?
5. Interior areas: will anything be moving while alarm is armed? Is the building heated and/or air conditioned?

Security Checklist

Doors. Knowing the location and construction of exterior doors is important, since the point of entry of most burglaries is a door. Protecting interior doors provides "trap" zones. If an intruder bypasses perimeter protection devices, he may open a protected interior door and trigger an alarm.

Doors with glass panes or all-glass doors should have glass-breakage detectors as well as contact switches. Metal doors require wide-gap magnetic contact switches to assure an adequate gap between the magnet and the switch.

Look for unused or boarded-up doors. They should be protected even though they are rarely used. I have known business owners who have forgotten about boarded-up doors. On one occasion, I discovered such a door while inspecting the outside of the building. It was covered on the inside by thin paneling and thus was not noted on my internal inspection.

Windows. Protecting windows is a debatable issue, one that you must resolve for yourself. There are positive aspects to both protecting and not protecting windows.

Fitting each window with foil tape or glass-breakage detectors (not vibration contacts) assures that entry by breaking the glass will be detected immediately. Foiling windows is a time-consuming process, however, and can quickly increase the price of an alarm system due to the labor involved. The tape is inexpensive, the labor is not.

Glass-breakage detectors that are placed directly on the glass have limitations if windows consist of several small panes. Aesthetics are important, too.

Audio discriminators—sound detectors—are becoming increasingly popular with alarm installers. One audio switch can protect several windows by listening for the sounds of breaking glass or forced entry.

If an intruder gains access through an unprotected window, he will most likely leave through a door, especially if he is carrying a large or heavy object. In that case, if all doors are protected, he would be detected leaving instead of entering, possibly minimizing the loss, but not preventing it.

An alternative to protecting each window is to make an interior (trap) zone. Installing concealed magnetic contact switches on one or more interior doors or using space-protection devices to cover areas where an intruder is likely to pass will reduce the time required to install the system and therefore reduce the cost.

Of course, it means that an intruder must be inside the building before he is detected.

Whether or not you protect every window will depend on the level of security you want to provide. Maybe protecting the most vulnerable ones will suffice; maybe protecting none is acceptable. (There may be some legal liabilities created by not protecting all potential points of entry, however. We discuss those liabilities in Chapter 19.)

Walls. Hollow walls can be used for concealing wire, solid walls cannot. If that seems obvious, it is, or at least it should be. Nevertheless, many installers have arrived at job sites with installation plans that did not take solid walls into consideration.

Space-protection devices should be mounted on vibration-free walls. Microwave detectors should not be aimed at walls through which their radiated energy will penetrate.

Common walls present an unusual and sometimes overlooked problem. If an intruder gained entry into an adjoining building or room, could he gain entry into your customer's building through a common wall? If so, you should protect the common wall with a photoelectric beam, vibration contacts, or other devices.

Floor/Ceiling/Roof. In addition to considering whether or not access could be gained through the floor, ceiling, or roof, consider the installation of the system. Is there enough room in the crawl space above the suspended ceiling or in the attic for the installer to work and pull wire? If not, wire may have to be exposed or wireless RF (radio frequency) equipment may have to be used.

If the roof has gable ends, siren speakers or bells may be mounted inside a vent. Soffit vents can be used for siren speakers if they are accessible.

Older buildings that have been remodeled may have a common crawl space above the ceiling. If so, trip wires or some other type of detection device may be needed in the common area.

Interior Areas. Microwave detectors are affected by fluorescent lights. Passive infrared, ultrasonic, and microwave detectors may be affected by air currents if they are located near a vent.

Will pets or anything else that moves, such as hanging mobiles or motorized displays, be in the building when the alarm is armed? Motion detectors detect movement regardless of what moves; passive infrared detectors sense heat changes, and animals give off body heat.

Remote-controlled televisions receive their commands by ultrasonic signals. Therefore if you are installing an ultrasonic detector in a room with a remote-controlled television, be sure to check for possible interference.

If the building's temperature drops below 40 degrees Farenheit (4.4 degrees Celsius) or goes over 120 degrees Farenheit (48.9 degrees Celsius), under normal circumstances, care should be taken with the types of detection devices used. Extreme temperatures, especially for prolonged periods, may affect batteries and/or other items of equipment.

Other Considerations. In addition to the preceding information, you may want to know about the business' operation if it is a commercial system, or about the family's lifestyle if it is residential. The more information you have, the better job you will do as a system designer.

If you are designing a residential system, does the homeowner want to use the alarm while at home or away, or both? If you are planning on installing space-protection devices, they will have to be turned off (shunted) while the family is in the house and the system is armed. Are neighbors close enough to hear a bell or siren? Would a strobe light pinpoint the location of the alarm for a passing patrol car? Can the system be expanded without major modification or replacing the control panel? Keep in mind that additional sales may be made if you do not install a complete burglary/fire system initially. Will you be able to add heat and smoke detectors, additional zones, or other devices later?

You're the Expert

When designing an alarm system it is important to consider the prospect's requests. But, you are the security expert. If you can accommodate the requests in your design, so much the better. If you cannot, explain to him the reasons why.

Many dealers have taken a firm stand on the types of protection they will provide. Anything less than total protection is just not acceptable to some, and they will not install a partial system. While others will install partial protection, they carefully outline in detail in the contract the areas that are not protected and the gaps that exist in the building's security.

Survey Form

Rather than provide a sample form for you to copy, you should design your own. Review the questions presented earlier and develop a survey form that will work for you and your company. You will probably want one for commercial systems and another for residential systems. Of course, you could make the forms even more specialized and have specific ones for retail stores, offices, apartments, and homes, or whatever types of installations you normally do.

When designing forms, leave enough space for the salesperson or system designer to answer the questions in detail. Space to include drawings and sketches will be helpful when you begin designing the system.

SELECTING EQUIPMENT

The second step in designing the system is determining what will be protected and what specific devices will be used. Careful analysis of the survey form and drawings will aid you in determining your equipment needs.

The third step is selecting the control unit, one that will accept the detectors in the system and meet the customer's requirements. There are many excellent control panels on the market. Using two or three that will accommodate 90% of your customers is ideal, if you can do it. If you use too many different types of control panels, it will be difficult for your installers and service personnel as they will not install any one control often enough to become familiar with its operation and service requirements.

Using too few control panels presents a different problem. If you only use one type, you may be customizing your customers to your systems instead of customizing your systems for your customers. Also, you may not have enough features to satisfy the needs of some customers.

Determining what the system will do is the fourth step, that is, after the alarm has tripped, what happens? A local alarm is one that just makes noise. But what kind of noise? A bell? A siren?

Personal preferences and local ordinances will determine whether you select bells or sirens. Both have appropriate applications. My personal choice is a siren. A good siren driver, powered by twelve volts or more, connected to a thirty-watt speaker can create an ear-splitting wail. Speakers may be easier to mount and they require less maintenance than bells, especially if they can be mounted inside attic vents.

As long as noise is being made outside the building, consider making noise inside, too. This can alert a sleeping family that an emergency exists. Even if no one is home, it will notify the intruder that his presence has been detected.

The primary function of an alarm system, in most cases, is to make noise and alert people to the presence of a problem. (Many of the new interior siren speakers are attractive devices, with one model having the appearance of a smoke detector.) Other annunciators include strobe lights and tape and digital dialers. They are frequently used in conjunction with bells or sirens. Some installations may require direct lines from the protected premises to alert an alarm company's central station.

A system that relies solely on nonaudible devices (silent alarms) has one potentially serious problem: it assumes that the police or other persons will respond immediately. This may not be a valid assumption, and a delayed response could mean a significant loss. If in doubt, make noise.

Specifying and Purchasing

The final step is to order the equipment. There are hundreds of security and fire equipment manufacturers and distributors. Generally, it is advisable for a small alarm company to use the services of a distributor. He has a wide variety of products in stock and in effect becomes an alarm equipment supermarket. Many distributors offer design assistance and some sponsor training seminars and trade shows. Locating several dependable distributors who offer good service and competitive prices can alleviate many of the problems associated with purchasing equipment.

Equipment manufacturers are an asset to the dealer/installer as well. Their technical staffs can assist with special applications or unusual problems that might develop during an installation. Most manufacturers have toll-free telephone numbers for dealers to place orders and request assistance.

NEW TECHNOLOGY

Because we have not discussed the various types of equipment yet, a section on new technology may seem out of place in a chapter on designing alarm systems; however, it is not. Later chapters deal with specific features and functions of alarm components. Right now, let us take a less technical and more philosophical approach to high-tech equipment. Specifically, let us explore some of the benefits of the new technology.

Integrated circuits and microprocessors are changing the nature of the alarm industry. The move from relays to transistors represented a significant advance in technology, and the new computerized alarms are an even greater step forward.

The benefits to the dealer and installer are increased reliability and function as well as reduced installation and service time. The benefits for the customer are additional features and greater security.

Salespeople, installers, and customers view the features of the new technology differently. Salespersons will translate the features into benefits for the customer and use them as a selling tool. Installers look at them from a technical standpoint; they want to know how simple—or difficult—the system's installation and service will be. To the customers, the features represent solutions to problems, means to achieve a certain level of protection.

Simplicity

Whether or not you are installing a high-tech alarm system, simplicity has advantages. Systems that are not simple for the customer to operate will mean extra work for you and needless service calls. Almost half of all false alarms are user caused. Making a system complex will increase that number.

You are designing systems for people who may not be as technically competent as you are. Most of my customers are not as gadget-oriented as I am and do not share my fascination with buttons, switches, beepers, and lights on a security system. So, keep the customer's needs as well as his operational abilities in mind as you design his system. Keep it user simple.

Educating the customer on his system's operation is important. Additional instruction time may be required initially with the high-tech systems, but it could reduce the need for return visits to reinstruct the customer. Also, it could reduce or eliminate service calls because of improper use of the system. Occasionally, it may be necessary to provide written instructions.

Installer Friendly

A system's installer friendliness is as important as its user friendliness. Systems that are difficult to install may become too labor intensive or they may never be installed correctly. Either way, profits will be eroded. Many of the newer systems are not difficult to install; some go in more easily than the older systems.

Such ease of installation is partly because of some new features; these features may simplify troubleshooting and servicing as well. Some of the newer systems even eliminate the need for an installer or technician to program certain functions into the control. Most of the information, such as entry and exit delay times and instant and delay zones, can be programmed into the system by a remote computer. After it is installed, the installer tells the central station what data are to be entered into the control; then, almost instantaneously, it is done.

Adding features to a standard high-tech alarm system may be as simple as plugging in a modular unit. Similarly, once a problem is isolated, servicing usually means simply replacing a plug-in module or circuit card. Thus, your customer will not have to wait several weeks for a component to be repaired or replaced.

High-tech systems provide more information to the user, dealer, installer, and troubleshooter. Detection devices and zones can be programmed in or out of the system. Circuits that are not normal, because they are open, shorted, or have some other problem, may announce their troubled status via lights or buzzers. Additionally, many systems include an arming inhibit feature that prevents the alarm system from being armed (turned on) if any circuit is in trouble. Loss of AC power of a low battery also may be annunciated locally or remotely.

The more information that is available, the greater your ability to provide reliable security to your customers and reduce service and maintenance expenses. While information is helpful to some, however, it may be confusing to others. But the more you have, the better you can serve your customers.

Choosing High-tech Equipment

Deciding which high-technology equipment is appropriate for your company to install requires careful planning. Changing products frequently, that is, trying a variety of new items as they become available, requires additional inventory, which ties up capital. Moreover, it means that installation and service personnel must be familiar with more types of products.

You might want to evaluate new equipment before installing it for a customer. Put the new item in your own home, or in a "friendly" installation, and test it before you install it in the field.

New products do not always meet dealers' expectations and sometimes they do not perform in accordance with manufacturers' claims. With any new product, problems can be expected, and alarm equipment is no different. Overall, the new products offer improved features, more reliability, and the potential for greater dealer profits. Many manufacturers are working closely with dealers to design products that meet their needs and are doing an excellent job, the few problems that exist notwithstanding. The complicated process of evaluating and selecting new high-tech equipment does not lend itself to a simple solution. Part of your job as a system designer is to evaluate the equipment that is available and select what is best for your customers.

Bidding the Job 10

Accurate bidding is crucial to the success of your alarm company. The market appears to be large enough to handle an industry many times its present size. Yet it is somewhat ironic that many alarm dealers will close their doors this year, bankrupt or nearly so.

The reasons for these failures are numerous, but they can be reduced to a common denominator: inability to make a profit large enough to sustain operations. Many other alarm companies, though they have not failed, will have financial problems. They, too, are not maximizing profits.

Personally, I suspect that the financial problems started when the sales rep handed the proposal (bid) to the prospective customer. There are two potentially significant threats to the bidding process that yield the same net result: zero profit (or worse, a loss). First, a bid that is too high in a competitive bid situation may not win the job and a customer and future referrals will be lost. Second, if the job is grossly underbid, the dealer may not be able to cover his expenses, much less earn a profit.

Of the two threats, the latter is more dangerous. If you are not going to make a profit or at least break even on an installation, you might as well keep your crew in the shop and have them polish the trucks. Occasionally, you can submit a bid above the lowest amount and still be awarded the contract.

As an alarm installation company, Sanger and Son, Inc., was rarely the low bidder. We were competitive, but not on the low end of the scale. Many of our customers have told us, after the bids were awarded, that we were not the low bidder, but we were selected because they were confident that we could install and maintain the systems professionally.

Sure, many of our customers could have had alarm systems installed for less money. But they chose to pay us because of our professional attitude and the confidence that we instilled in them.

Except for rare instances, it was never the goal of Sanger and Son to be the low bidder in competitive bidding. Of course, there were a few times when we wanted an account for a particular reason, such as for referrals or maintenance on other systems. So, even though we might not have made a profit on the installation in those few instances, we earned more from referrals or in other ways.

Even with the best of bidding practices, there may be a few installations that will be unprofitable. Occasionally, some jobs will be more than you had anticipated—it is almost unavoidable. Fortunately, by following some basic guidelines and developing good bidding practices, the unprofitable jobs will be kept to a minimum.

Assuming that the sales rep has done his job adequately and provided the system designer with accurate data, the next step is to prepare a proposal for the prospect. To do that, you will need to know the types and quantities of equipment that will be used and how much labor will be required to install it. Developing an equipment list is relatively simple; estimating man-hour requirements is not. As a guide, a functionally effective activity time (FEAT) is provided in Table 10.1.

Table 10.1 Functionally Effective Activity Time Chart

Equipment	*Man-hours (Hard-wired)*	*Your Estimate*	*Man-hours (Wireless)*	*Your Estimate*
Control panel (surface mounted)	2.00	———	1.50	———
Control panel (recess mounted)	3.00	———	2.00	———
Transformer/power supply	0.50	———	N/A	N/A
Siren driver	0.25	———	N/A	N/A
Zone board	0.50	———	N/A	N/A
Remote station, key (recess mounted)	1.00	———	N/A	N/A
Remote station, digital (recess mounted)	1.50	———	0.50	———
Remote annunciator (recess mounted)	1.50	———	N/A	N/A
Digital communicator or tape dialer	0.75	———	N/A	———
Contact switch (surface mounted)	0.75	———	0.20	———
Contact switch (recess mounted)	1.00	———	0.30	———
Overhead door contact	1.50	———	0.75	———
Glass-breakage detector	0.50	———	0.25	———
Foil tape (per unit, window or door)	0.75	———	0.50	———
Space-protection device (surface mounted)	1.00	———	0.25	———

Table 10.1 (*continued*)

Equipment	*Man-hours (Hard-wired)*	*Your Estimate*	*Man-hours (Wireless)*	*Your Estimate*
Space-protection device (recess mounted)	1.50	______	0.75	______
Panic/hold-up button (fixed location)	1.00	______	0.30	______
Heat detector	0.75	______	0.25	______
Smoke detector	1.00	______	0.30	______
Bell and box	2.00	______	N/A	N/A
Siren speaker, exterior	1.50	______	N/A	N/A
Siren speaker, interior (surface mounted)	1.00	______	N/A	N/A
Siren speaker, interior (recess mounted)	1.50	______	N/A	N/A
Wireless transmitter	N/A	N/A	0.10	______
Wireless receiver	N/A	N/A	0.50	______
Experience factors for installers				
Unskilled/trainee	2.1			
Apprentice	1.5			
Journeyman	1.0			
Skilled	0.6			

Note: an informal survey of alarm dealers across the country revealed that an "average" installer was considered to be at the journeyman level, and is the basis for thé FEAT times. Installation times for three other proficiency levels, unskilled trainee, apprentice, and skilled installers also were tabulated. The installation times for these three levels varied by 210%, 150%, and 60%, respectively. The experience factors shown in the chart reflect these variances.

FEAT CHART

Keep in mind that the FEAT chart is not a magic formula, it is a guide. Several variables must be considered: the experience of installers (or installation crews), the exact type of equipment being installed, and the type of structure being protected.

In its present form, the FEAT chart may not list all of the equipment that you install, but it will give you a start in preparing your own. Make a list of the items of equipment that you use that are not included in Table 10.1, then estimate the time it would take an average installer to install them. Better yet, if you can observe your installers at work, time their activities, and take the average of those times, you will have a very good estimation for the chart.

Table 10.1 shows the times required for an "average" installer to install certain types of equipment. Use it as a guide until you can develop one that fits your organization. Chances are that your chart will contain different times if you are installing specialized equipment or if your market has special requirements.

Using the equipment list prepared by the system designer, calculate the estimated man-hours for an average technician to install the items. (Be sure that the times you enter are for your average installer. Experience factors are included at the bottom of the chart ranging in experience from trainees to skilled installers.) Then multiply the total number of hours by the appropriate experience factor. Table 10.2 shows a hypothetical system's equipment and estimated man-hour requirements.

Estimating Equipment and Labor Requirements

The equipment for the hypothetical system would cost $851.00 and it would take 18.75 man-hours to install. The estimate is based on an average installer; if a less experienced installer were assigned to the job, it would take longer. An apprentice might require 28.1 man-hours (18.75 man-hours × 1.5 experience factor) and an unskilled/trainee installer might require more than 39 (18.75 man-hours × 2.1 experience factor). If an experienced, skilled installer is assigned to the job, it might take only 11.25 man-hours (18.75 man-hours × 0.6 experience factor).

Installation hardware, supplies, and wire have not been included in the cost estimate, nor have travel, set-up, and clean-up time. How you handle these costs will depend on how your business is organized. They could be included as part of overhead or you could attempt to estimate how many feet of wire and how many screws you would use for a particular job. It may be simpler, and just about as accurate, however, to calculate supplies and hardware as a percentage of the total equipment cost. Usually, 5% of the total is adequate to cover miscellaneous items. In this case, 5% of $851.00 is $42.55, which is probably adequate.

Including travel, set-up, and clean-up time is a bit more complex, but it is

Table 10.2 Estimated Equipment and Labor Requirements for a Hypothetical Alarm System

Item	*Quantity*	*Unit Cost ($)*	*Extended Cost ($)*	*Unit Man-hours*	*Extended Man-hours*
Control panel (recess mounted, hard-wired)	1	150.00	150.00	3.00	3.00
Power supply	1	40.00	40.00	0.50	0.50
Siren driver	1	20.00	20.00	0.25	0.25
Zone board	1	60.00	60.00	0.50	0.50
Remote annunciator (recess mounted)	2	20.00	40.00	1.50	3.00
Wireless receiver	1	40.00	40.00	0.50	0.50
Digital communicator	1	50.00	50.00	0.75	0.75
Contact switches (surface mounted, hard-wired)	3	4.00	12.00	0.75	2.25
Contact switches (recess mounted, hard-wired)	3	4.00	12.00	1.00	3.00
Contact switches (surface mounted, wireless)	3	4.00	12.00	0.20	0.60
Wireless transmitter (for switches above)	3	20.00	60.00	0.10	0.30
Space-protection device (surface mounted, wireless)	2	80.00	160.00	0.25	0.50
Portable panic button (wireless)	1	20.00	20.00	N/A	N/A
Smoke detector (wireless)	2	60.00	120.00	0.30	0.60
Siren speaker, exterior	1	30.00	30.00	1.50	1.50
Siren speaker, interior (recess mounted)	1	25.00	25.00	1.50	1.50
Totals			851.00		18.75

Table 10.3 Equipment Cost and Labor Estimates

Factor	*Equipment Cost ($)*	*Labor Man-hours*
Primary equipment	851.00	18.75
Wire, hardware, supplies	42.55	
Travel, set-up, clean-up	———	3.00
Total equipment cost	893.55	
Total labor cost*	217.50*	21.75*
Total installation cost	1111.05	

* Installers pay = $10/hour.

not difficult to calculate. A series of simple steps will provide a good estimate of total time requirements:

- First, divide the estimated installation time by the number of installers: e.g., 18.75 ÷ 2 = 9.38.
- Second, divide the number obtained in the first step by 6: 9.38 ÷ 6 = 1.56.
- Third, round the number obtained in the second step to the next whole (higher) digit: 1.56 rounded = 2.
- Fourth, multiply the number obtained in the third step by 1.5: 2 × 1.5 = 3.
- Fifth, add the number obtained in the fourth step to the installation time (man-hours) estimate from your equipment and labor requirements list (from Table 10.2): 3 + 18.75 = 21.75.

In the first step, we determined how many installers would be used and how many man-hours would be required with that number (9.38 man-hours with two installers). In the second step we divided by six, the number of productive hours in an eight-hour work day, to find out how many days the installers would be at the job site—1.56. Step three rounded the fractional days to whole days (two days). In the fourth step, we multiplied 2 days by 1.5 hours, the estimated time for travel, set-up, and clean-up per day, for a total of 3 hours. In the fifth step, we added the 3 hours to the estimated installation time of 18.75 man-hours for a total job time of 21.75 man-hours.

Although this method is not exact, it is accurate enough to provide you with a realistic time estimate for bidding a job. Table 10.3 shows the estimated equipment, labor, and total installation costs.

OVERHEAD

Overhead, or operating expenses, must be considered before we have an accurate picture of the true cost of the installation. Overhead consists of expenses such as rent, legal fees, insurance, utilities, vehicle maintenance, and salaries not covered by the installation charges. All of these expenses must be reflected in the price charged to the customer.

Let us assume that our small alarm company has operating expenses that total $2,000 per month. Our task becomes finding a method to pro-rate the overhead expenses among the installations completed during the month. If you estimate that you will be installing ten systems per month, some simple arithmetic tells you that $200 added to each system will cover the overhead expenses. What if some of those ten systems are large and some are small, however, and not the "average" that this method supposes? It becomes less than equitable. Also, if you only have one installation (since you are just getting started in business) this month, are you going to pass the entire $2,000 along to one customer? Of course not.

Determining overhead percentages is more easily accomplished in retrospect. That is, you need to have several months (or years) of operating experience to predict overhead expenses accurately and project them into bids. Once you have historical data about your expenses and sales, calculating and projecting overhead becomes simpler.

This is another one of those areas in which there is no substitute for experience and good records. Overhead expenses can be projected as a percentage of sales or as a percentage of installation costs; often the latter is easier to calculate.

For purposes of illustration, let us assume that for the past several months our company's total installation costs (equipment and labor) have been averaging about $15,000 per month and overhead expenses have been about $2,000 per month. Dividing $2,000 by $15,000 yields 0.1333 or 13.33%. In other words, overhead is 13.33% of what is being spent for total installation costs. Table 10.4 shows the addition of overhead expenses to hypothetical alarm system installation costs. The actual cost of the system is not $1,111.05 as we might have thought, but $1,259.15 after overhead is added.

All of the calculations we have done thus far have just brought us to a break-even point. So far, we have not considered the aspect of the bid that will sustain our business—profit.

Table 10.4 Equipment, Labor, and Overhead Cost Estimates

Item	*Cost ($)*	*Total Cost ($)*
Equipment	851.00	
Wire, hardware, supplies	42.55	
Total equipment cost		893.55
Total labor cost		217.50
Total installation cost		1,111.05
Overhead (13.33%)		148.10
Total system cost		1,259.15

PROFIT

It is up to you what you will charge for your goods and services. No one can tell you what to charge or how much profit to make. The only constraint on profits and prices that are too high is the fact that you will lose sales. If they are too low, or nonexistent, you cannot sustain operations.

Sanger and Son, Inc., uses a procedure similar to the hypothetical problem that was just presented. An Atari 800 microcomputer performs the calculations for us. I have known dealers who used different methods for calculating bids. Some were sophisticated and some were very simple. One dealer used to total the cost of equipment, double it, and add $20 per opening (door or window). It seemed to work for him, but it did not tell him how his costs were distributed and how overhead affected his profits.

If you are still wondering about the hypothetical alarm system and how much the customer would be charged, the only information I can provide is that if twenty dealers took the cost data, they would arrive at twenty different bid prices. Other than that, all I can tell you is how Sanger and Son, Inc., would bid the job. Figures 10.1 and 10.2 show bid prices for selling and leasing the system.

SELL OR LEASE?

Whether to sell or lease alarm systems has been the subject of controversy for many years. Selling systems provides good cash flow and an up-front profit; leasing provides recurring revenue. Many alarm dealers do both, depending on their customers' needs. Depending on your company's organization and philosophy, you may decide to sell, lease, or do both. It is a decision that requires careful attention and frequent review.

The only complaint that I have, personally, involves leasing. What some dealers call "leasing" is not. They calculate the cost of equipment, labor, and

Equipment ($893.55 + 50% mark-up)		$1,340.33
Labor (21.75 M/H × $30/hr)	$652.50	
Overhead	148.10	
Installation charge		800.60
Bid price (sale of system)		$2,140.93
Less total system cost		(1,259.15)
Less sales commissions*		(321.14)
Gross profit		$ 560.64
Monitoring = $17.50 per month.		

* Sales commissions total 15%. That is, sales rep and sales manager share a commission based on the amount of participation each contributes to the sale. Typically, with no assistance from the sales manager, the sales rep earns a standard 12½% commission on the bid price.

Figure 10.1 Selling price for a hypothetical alarm system.

Labor (21.75 M/H × $30/hr.)		$652.50	
Overhead		148.10	
Installation charge (lease)			$800.60
Monthly lease fee	= $21.66*		
Monthly monitoring fee =	17.50		
Total monthly charges	$39.16		
Installation charge		$ 800.60	
Less system cost		(1,259.15)	
Loss on installation		($ 458.55)*	

* Monthly lease fee is based on a thirty-six-month payback period for the loss on installation and commissions:

Loss	$458.55
Commissions	321.14
Total	$779.69 ÷ 36 = $21.66

(Sales commissions, paid monthly, continue as long as the account is active, so sales reps can earn more if the system is leased.)

Figure 10.2 Lease price for a hypothetical alarm system.

overhead, and then add on an amount for profit. This becomes their "installation" cost for a leased system, plus a monthly lease fee.

Applying the same philosophy to car leasing probably would put a company out of business. Would you pay the full price for an automobile when it was delivered to you, then pay a monthly lease fee? That is like paying for the car in full, up front, and then paying a certain amount each month for the privilege of driving it.

Under a lease agreement, an alarm dealer could cover part of his costs and charge a monthly fee. The benefits of leasing are: (1) the customer gets an affordable alarm system and (2) the dealer retains title to the equipment and earns a monthly income in the process.

Following Up/Closing the Sale 11

As discussed in Chapter 8, I prefer a two-call sales approach. On the first call I obtain information about the prospect. With that information I can design a system, prepare a proposal, and make a second appointment to close the sale.

OBJECTIONS

When you make the second call to close the sale, you will be presenting your suggestions to the prospect. He may have some questions about the system and he may have some objections. These are an integral part of the sales process. Once you accept that fact, you will be better able to handle them. Beginning salespersons may feel disheartened when objections are raised. They may look like insurmountable obstacles that will prevent a sale. Objections need not be that discouraging.

Veteran salespeople welcome expressed objections because they can be turned into sales aids. The hardest type of prospect to sell is the one who does not say anything or show any interest (or disinterest) in your proposal. Similarly, a prospect who agrees with everything is difficult to sell because you may be unable to determine what he is really thinking.

Honest objections tell you how far away you are from closing a deal. They are raised for two reasons: (1) the prospect does not understand what you have told him or (2) he does not have enough information to grasp the significance of your point. You should consider objections as requests for more facts.

Objection-Handling Strategies

Handling objections successfully requires you to develop certain attitudes and methods (many of which can be applied to numerous human relationships outside the selling process).

Avoid Arguments. Getting into an argument with a prospect is one of the easiest and most disastrous things you can do. Your position with him is one of cooperation, not one of conflict. Few people are really convinced of anything by argument.

Don't argue—suggest. Allow the prospect to draw his own conclusions based on the facts you have presented. The old adage, "Win an argument, lose a friend," is appropriate in these instances. Except, that it might be worded: "Win an argument, lose a sale."

Work on anti-argument phrases. If a prospect objects to price by saying, "Your price is too high," do not counter by replying, "It's high, but it's worth it." Instead, say something like "Yes, it is expensive. But many of our customers have said it is the best investment they've ever made."

Listen to the prospect's objection and try to find some point with which you can agree. Then restate this point of agreement and proceed with your closing.

Irrelevant Objections. Sometimes the prospect will object to some point that is unrelated to what you are selling. Unless the objection directly deals with the proposition at hand, you will be wise not to take issue with his statement.

Answering Objections. Usually, it is a good idea to answer an objection as soon as it is raised. Failure to do so might lead the prospect to think that you do not have an answer and hope that it will be forgotten. Or he may be concentrating on the objection, thinking that you will not come back to answer it, so that he misses the rest of what you are telling him.

A few instances exist when you will want to postpone responding, for example:

1. When a price objection arises early in the close
2. When the objection will be answered more effectively later in the presentation
3. When the objections are so frequent and so petty that you think they are simply an effort to slow down the presentation

Before answering an objection, listen carefully to what the prospect is saying. Let him state himself fully. Even though you have heard similar comments many times, act interested.

Objections to Price

The most frequently voiced objection is that the price is too high. You will have to determine if the prospect really believes that or if it is easiest objection to verbalize. If he truly considers the price to be too high, he may mean one of two things: first, in his opinion, the alarm system is not worth the price you have set or second, he does not have enough money to pay what you are asking. Naturally, the approach used in handling the first objection would be different from that for the second. In the latter case you could attempt to work out some kind of payment plan. In the former, you would have to build up the value of the system in the prospect's mind.

THE CLOSE

Everything that you have done so far, from prospecting to handling objections, is focused on one final event: the close. The close, simply stated, is nothing more than getting the prospect's agreement to a sales contract: he agrees to buy what you are selling.

Good salespeople are good closers. Poor closers are poor salespeople. The salesperson who cannot close is like the runner who trains for a race, gets off to a good lead at the starting gun, then falls flat on his face ten yards from the finish line. By the way, I have heard people say, "I'm a good salesman, but a poor closer." It is a contradiction in terms. You cannot be a good salesperson if you cannot close sales.

An old story that circulates between professional salespeople goes something like this: A novice salesperson reports to his sales manager and says, "I could lead 'em right up to the watering trough, but I couldn't make 'em drink." His boss looked up and said, "Drink? Whoever said you had to make them *drink?* It is your job to make them *thirsty!*"

The fear of failure is probably the most common reason for failing to close a sale. When a prospect says "no," some salespeople take it as a personal rejection. Hearing the word does not hurt; there is no pain involved, except a bruised ego. You can learn to ignore the word. You can even develop a deafness for it. Even if the final answer is an absolute "no," you should be able to bounce back and go on to the next prospect. If you can't, you should start looking for a job in a field other than selling.

Basic Closing Tactics

Several specific closing techniques are discussed later in the chapter. Here are some general ones that you may find helpful.

Isolate the Prospect. Closing difficulty increases when uninvited third parties appear. If a business associate or family member drops in during your close, he may bring additional objections and distract the prospect.

When you are ready to close the deal, try to isolate your prospect to minimize interruptions. Remember the last time you bought a car? You were probably taken to a small office (closing booth) to make the final arrangements. Many sales can be closed at private dinners or clubs, if you have arranged for privacy.

Your Attitude. As the close approaches, you are likely to feel increased nervous tension. This is natural, but you should be careful to maintain a calm, casual attitude. The potential buyer may think that the nervous excitement is the sign of a novice salesperson; while this is not so, he may rationalize that if you are so excited over making a sale, you probably do not make many. He may carry this inaccurate logic a step further and reason that if you do not make many sales there may be a good reason—one that he has overlooked—and he will delay the sale until he can think it over a bit longer.

Approach this step with confidence, as though you close several sales a day and it is all in a day's work. Confidence is contagious. If you display confidence, your prospect will have confidence in you. If you doubt the outcome, your prospect will share that doubt. A professional confidence should flow from you to your prospect, creating a relaxed atmosphere that should carry through to your departure and through the installation of the system.

Simplify the Contract. Your contract should be as simple and short as possible. People are wary of lengthy contracts and fear that hidden somewhere in the fine print are legal gimmicks to protect you and leave them at your mercy.

Because the nature of the alarm business is such that lengthy contracts are not uncommon, you may find it advisable to go over all of the provisions with the buyer and make certain that he understands them fully. Whenever possible, use common language instead of legalese. The buyer also will be impressed with the seriousness of the contract. If he is made to feel that he has certain important obligations to match your obligations, he is more likely to agree to the contract's terms.

If your prospect hesitates at the actual signing, you should point out that the contract is binding on your company as well as the buyer. You can show him where it stipulates the products and services that you will be bound to provide. Some salespersons use linguistic devices. For example, they avoid using the word "sign," relying instead on statements such as: "If you will just *OK* this agreement," while handing the contract to the prospect.

Some bring out the contract at the beginning of the closing session so the prospect will become accustomed to seeing it. It also suggests the idea that it will be used soon.

Closing Time. Many learned, professional salespersons have written books and articles on the right time to close a sale. It is doubtful that any single "right" time exists; probably there will be many opportunities.

Seasoned salespersons recognize signs of a prospect's increased interest. Some of these are voluntary, some are involuntary. For example, a voluntary sign might be a question from your prospect: "How soon could the system be installed?" Many top salespeople believe that when a prospect asks the price, he is very interested, and when he inquires about particular terms, he is practically sold. These are voluntary signs and are fairly easy to recognize.

You must pay particular attention to the involuntary signs, however. The prospect's apparent indifference fades away and is replaced by interest. He may lean forward in his chair or his eyes may project less skepticism. Closed hands may indicate that he is not yet convinced; open hands reflect a more relaxed posture and willingness to accept the proposal. Rules for recognizing involuntary signs do not exist. You will have to learn to recognize them through experience.

"No" Does Not Mean No. Good salespeople do not expect a "yes" answer the first time they ask the prospect to buy. Often it will take several closing attempts. You may be turned down several times before you make the sale, so do not give up after the first "no." Build your sales presentation around the expectation of being turned down. Each time you are turned down, add more value to the presentation and try again.

Another favorite rule of professional salespersons is the ABC rule—Always Be Closing. It is not quite that simple, but it is a good idea.

Reserve Punch. When you are going through your sales presentation, save a few strong points for the close. You do not want to run out of ammunition just when you need it most.

If you see that your prospect has almost reached a decision but is still hesitating, bring up the big guns. Call his attention to benefits that you have not yet mentioned or have touched on only briefly. Often this is enough to close the sale.

Closing Methods

There are several specific tactics for assisting your prospect to make a final decision. Many of them are used together. It is doubtful that you will be able to sell successfully without them.

Positive Replies. Throughout the interview when closing the sale, ask the prospect questions that cannot easily be answered negatively. Try to get a positive response to all questions. The theory is that you should try to get your prospect to think positively. That is, it attempts to make the closing decision only another favorable decision out of many made during the presentation. You are gently leading the buyer to the desired point by allowing him to build an affirmative attitude.

Assuming the Sale. Throughout your first and second meetings with the prospect, you should be planning on making the sale. You should be saying to yourself, "He is going to buy. He is a good prospect. He has a need for my products and services and can afford to pay for them. There is no reason why he will not buy. He is going to buy." These positive thoughts should be reflected in your attitude.

If you used this assumptive attitude throughout the presentations, it should not surprise your prospect when you use it during the close. As you explain features and benefits, say "when" instead of "if" and "will" instead of "would." For example, "You *will* feel more secure and have more peace of mind *when* we install your security system."

Minor Points. Another technique is to shift from major points to minor ones. It is often simpler for prospects to make a series of minor decisions instead of one major one. You might ask, "Are you interested in our lease plan or are you thinking of outright ownership?" This focuses the prospect's attention on a relatively minor point instead of on making a major acquisition decision.

Narrowing Choices. Many salespeople are prepared to present three proposals for alarm systems when they return to visit the prospect. One is a top-of-the-line system, another is a moderate version, and the third affords only basic protection. After determining where the prospect's interest lies, the choice should be narrowed to two.

Summarize Important Points. Lawyers summarize their arguments, helping the jury remember the many bits and pieces of evidence on which their case is based. Salespersons should do the same. Sum up the primary points, emphasizing those in which the prospect has shown an interest.

Overlooking the Obvious. Some salespersons have well-planned and executed sales presentations, yet never ask for the order. If you want to make the sale, you have to ask the prospect to buy.

Few would ask bluntly, "Do you want to buy?" Instead, most prefer to word the question along the lines of the assumptive close, using such questions as; "Will you be paying the deposit now, or shall we bill you?" or "Will you look over this agreement and OK it right here" (pointing to the place for the prospect's signature).

FOLLOWING THROUGH

You followed up on your first visit with your prospect by preparing a proposal and delivering it on a second visit. Now that the second, or closing, visit is completed, what is next? Following through with the sale.

The purpose is to make certain that the buyer is completely satisfied with his purchase. For the alarm salesperson, the follow-through relationship may last for many years.

When a system is sold (or leased, for that matter), it is important for the buyer to be educated on how to obtain the maximum benefit from it. The price he paid probably included a margin for this service, and he is entitled to it.

Another factor is the likelihood of repeat business or referrals. If a long-term relationship is hoped for, with frequent referrals from the new customer, he should receive periodic calls or visits.

To determine whether you should devote a given hour to selling or follow-through efforts, you should evaluate which activity will result in more sales in the long run, because the basic purpose of any follow-up is to increase sales. You are not giving away your time; you are investing it to earn greater sales.

APPLICATIONS III

Basics 12

Compared to other industries, the basics for the alarm industry truly are relatively basic. That is, you can enter the industry without being an electronics technician or a licensed electrician. Some states and municipalities have licensing requirements, but most are more registration oriented than competence oriented. The state or local governments want to know who and where the alarm companies are, not necessarily that they do a good job.

Eventually, we will probably see more licensing requirements and more competence testing to obtain a license. Licensing is not necessarily bad if the legislation is drafted properly. Doctors, lawyers, and barbers have to be licensed, so why not alarm dealers? Our business is protecting lives and property, and the consumer should have some assurance that we know what we are doing. For now, though, such requirements are minimal.

Understanding the basics will be helpful when competence-based licensing is effected. The remainder of this chapter discusses some of the basics of electronic circuits, tools and equipment, and power supplies. For the novice, this chapter will serve as a guide, offering suggestions and encouraging further study. For the veteran, it can serve as a refresher course.

ELECTRONIC CIRCUITS

Few alarm dealers perform major repairs on system components. Most equipment is designed so that circuit boards, cards, or modules can be removed and replaced easily and the defective units returned to the factory for repair. Occasionally, you may have a need to do some minor repairs, however, and the variety of components used in electronic alarm equipment can be overwhelming.

Learning how to perform minor repairs successfully is far from difficult. All it really requires is that you learn how to identify components by their physical shapes and schematic symbols. You should be able to follow schematic diagrams, master the use of a few basic tools, and know how to wire and solder. Your first task is to learn to identify different types of components.

Power Supplies

In a general sense, power supply means more than supplying 117 volts of alternating current (VAC) to a control panel. Because 117 VAC is too much power for alarm systems, it must be stepped down by a transformer. The primary power source for most alarm systems is 12 or 24 VAC, with a few operating on 16.5 or 18 VAC. Typically, burglar alarms use 12 VAC and fire alarms use 24 VAC.

A transformer is an inductor. Actually, it consists of two or more inductors located physically close to each other so that current passing through one will induce a current in the others.

Alarm systems are used 24 hours a day, 365 days a year, and require constant power even if AC power should fail. Standby power in the form of batteries is a must for all alarm systems to prevent a power outage from disabling the alarm.

Batteries. Batteries are available in a variety of shapes, capacities, and chemical compositions. Dry cell batteries are familiar to most people because they are used in lanterns and portable radios. They are still used in some alarm systems, but their drawback is that they must be replaced regularly.

Standby power for most alarm systems is supplied by gelled-electrolyte or lead-acid batteries. Both types are rechargeable and are available in a variety of capacities for long standby times.

Switches and Relays

Switches are useful for directing voltage or signals to where they are wanted. They are used to turn power on and off. Also, some types are used as detection devices in protective circuits, which are discussed later. The principles of operation, whether in a protective circuit or as a separate device, are similar, however.

In its simplest form, a switch is an on/off (make/break) device used to connect or disconnect power to a circuit. Elaborate switches can perform several tasks simultaneously. Figure 12.1 shows some common switches and relay contacts.

To understand how switches operate, it is important to be familiar with two terms: pole and throw. They are used to define switching arrangements.

The number of poles for a switch refers to the number of switching positions available. The number of throws indicates how many elements within a switch can be physically moved from pole to pole. A single-pole–single-throw switch, for example, has one pole and moves (throws) one element from pole to pole.

Sometimes you might want to switch a high-power circuit using only a low-power switch, such as in an alarm system. If the switch is rated at only 1 amp and you want to operate a bell or siren requiring 2 amps, you could add a high-current handling relay. You also could use a low-current relay to trip the high-current relay and route power through the high-current relay's contacts.

Relays are special types of switches in which the contacts are moved magnetically. Passing current through the relay's coil creates a magnetic field, moving the spring-loaded contact assembly, or armature. Most relays used in alarm systems have normally open (NO) and normally closed (NC) contacts. When no current is applied to the relay's coil, one set of contacts is open, the other is closed. Applying power reverses the contact arrangement.

Resistors

As noted earlier, few alarm dealers perform major equipment repairs. Occasionally, you may find it necessary to replace some common components, such as resistors.

Two types of resistors are in general use: fixed and variable (Figure 12.2).

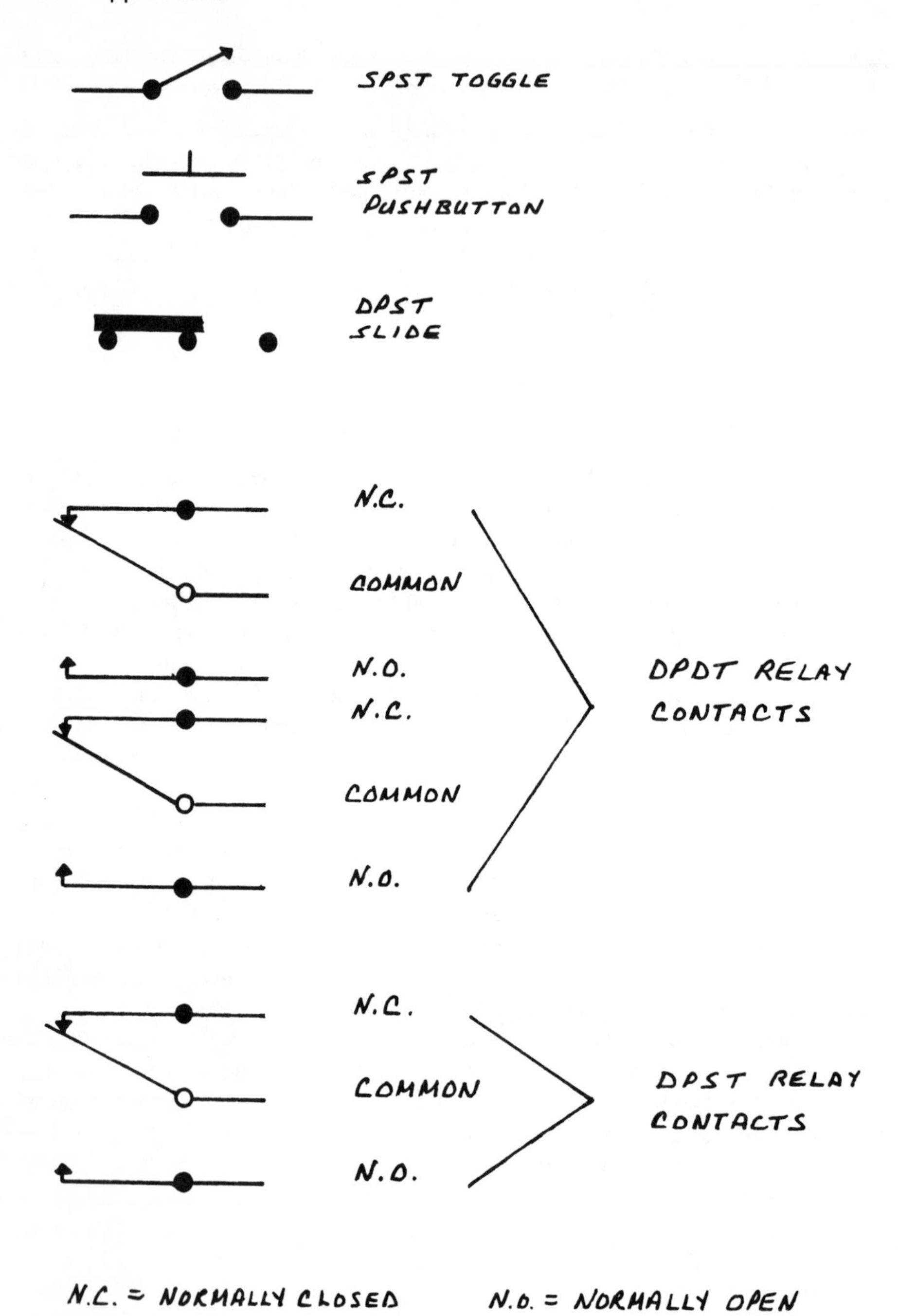

Figure 12.1 Switches and relays are common alarm system components. The diagrams show several types of switches and relay contacts.

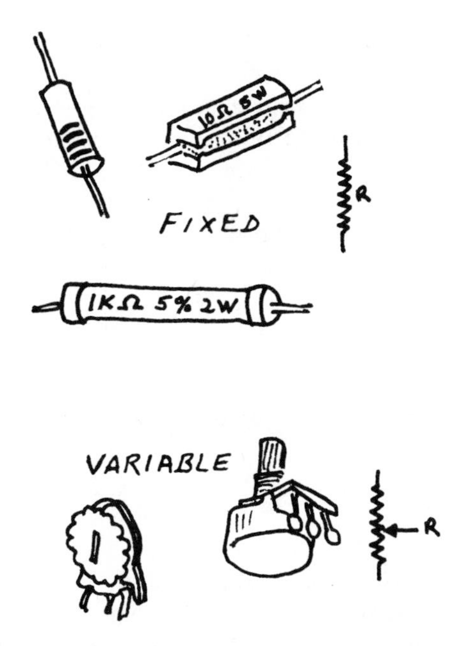

Figure 12.2 Fixed and variable resistors are available in a wide variety of shapes, sizes, and resistance values. Learning to recognize the common types will aid you in your security electronics work.

The fixed resistor usually has a cylindrical body with wire leads extending from both ends.

Carbon-composition resistors are common, ranging from a few ohms to several million ohms. The unit of resistance is the *ohm* (named after George Simon Ohm) and is represented by the Greek upper-case Omega, or Ω.

Expressing resistance values, especially if they are large numbers, can be confusing. Two prefixes are commonly used: kilo and mega. Kilo, abbreviated k, stands for thousand. Any number with a k following it, such as 2.5 k (read as 2.5 kilohms), must be multiplied by 1,000 to express its value. Hence, $2.5\ k = 2.5 \times 1{,}000 = 2{,}500$ ohms.

For resistance notations in the millions, mega is used, abbreviated M. A resistance notation of 1.3 M (1.3 megohms) would have a value of 1,300,000 ohms. That is, $1.3 \times 1{,}000{,}000 = 1{,}300{,}000$.

Most common resistors have no numerical markings to tell you their values. Instead, they are color coded. The code appears as different-colored bands around the body of the resistor, grouped nearer one end than the other. Holding the resistor so that the banded end is on the left, you can read the color code. The left band is band 1, the next is band 2, and so forth. Resistors have a minimum of three bands.

Bands 1 and 2 identify the two significant figures. Band 3 indicates the multiplier. The fourth band tells you the resistor's tolerance in percentage. If no fourth band is present, the resistor's tolerance is ± 20%. Table 12.1 shows the color code chart for resistors.

Listed below are several examples, showing you how the color code chart is used to determine resistors' values.

Example	Band 1	Band 2	Band 3	Band 4
A	Green	Brown	Orange	Gold
	5	1	1,000	±5%
B	Yellow	White	Brown	Silver
	4	9	10	±10%
C	Red	Black	Black	No color
	2	0	1	±20%

In example A, the value is 51,000 ohms (51 k), from 51 × 1,000, with a tolerance of ± 5%. Therefore the actual value can range from 48,450 to 53,550 ohms. In example B, the resistor would have a value of 490 ohms (49 × 10), ± 10% tolerance, for an actual resistance range of 441 to 539 ohms. In example C, the resistor's value is 20 ohms (20 × 1), ± 20% tolerance, or a range of 16 to 24 ohms.

You should note that a resistor's value has little to do with its size. A 5-ohm resistor can be similar in size or shape to a 3.3-M-ohm resistor. Usually,

Table 12.1 Resistor Color Code Chart

Color	*Significant Figure*	*Multiplier*	*Tolerance (%)*
Black	0	1	—
Brown	1	10	—
Red	2	100	—
Orange	3	1,000	—
Yellow	4	10,000	—
Green	5	100,000	—
Blue	6	1,000,000	—
Violet	7	10,000,000	—
Gray	8	*	—
White	9	*	—
Gold	—	0.1	5
Silver	—	0.01	10
No color	—	—	20

* Not used except in rare instances.

Resistor values can be determined easily using the color code chart.

the larger the resistor's package, the greater its power-handling capability. Resistors are not usually marked with their power-handling capabilities, which are usually ⅛, ¼, ½, 1, and 2 watts for carbon-composition types, so you will have to check the box or package when you purchase them.

The schematic representation for a fixed resistor resembles a saw edge, as shown in Figure 12.2. Variable resistors, also shown in Figure 12.2, have a similar schematic representation, except that an arrow is added. Most of the variable resistors used in alarm circuits are potentiometers, or "pots."

Capacitors

The basic task of a capacitor is to hold an electrical charge. How long it holds a charge depends on many things and is a measure of its tolerance. Unlike a resistor, which can be used at almost any voltage, a capacitor is designed to be used at or below a specific voltage level. Figure 12.3 shows the most commonly used fixed capacitors. They are divided into two basic groups: polarized and nonpolarized.

Nonpolarized capacitors can be installed in a circuit in either direction, without regard to polarity. The schematic symbol for nonpolarized capacitors is a straight and curved line.

Polarized capacitors must be installed in circuits in the proper direction. Polarity is important; most polarized capacitors indicate this in some way on their packages. The schematic symbol for polarized capacitors shows a + for identification (Figure 12.3).

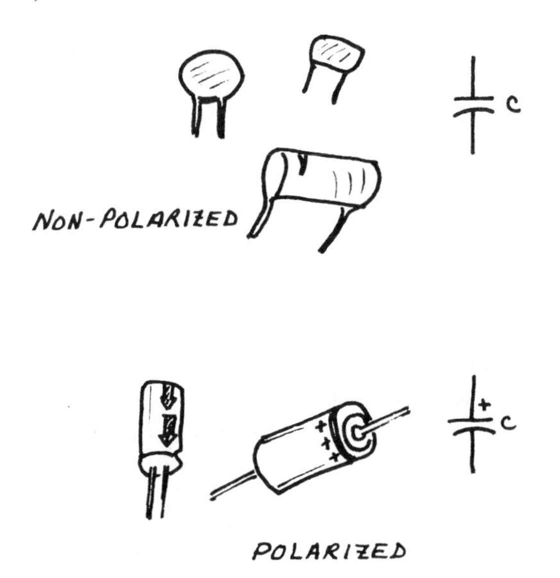

Figure 12.3 Capacitors are available polarized and nonpolarized. Be certain you use the correct type when replacing capacitors in electronic circuits.

Although variable capacitors are available, few are used in alarm system circuits that might require service by a technician. If variable capacitors require replacement or repair, it may be advisable to return the equipment to the factory.

The basic unit of capacitance is the farad (named after Michael Faraday) and is represented by the capital letter F. Most, if not all, capacitors used in alarm equipment have capacitance ratings much smaller than a farad. Two prefixes are commonly used to denote the smaller units—micro and pico, represented by μ (lower-case Greek mu) and p, respectively.

Micro indicates that a value is to be divided by one million to yield its value in farads. For pico, it is one trillion. Therefore 1.5 μF and 470 pF become 0.0000015 and 0.00000000047 F, respectively. Working with numbers such as these can become very cumbersome, so you can see why the shorthand notations are useful.

Most capacitors have values printed on them as well as voltage specifications expressed as WVDC (working volts DC). The WVDC specifies the maximum voltage to which the capacitor should be subjected. When replacing capacitors, you can use one with a higher WVDC, but not one with a lower WVDC rating.

Sometimes the type of capacitor is critical, and any substitutions will have a direct effect on how the circuit performs. If in doubt, consult the equipment manufacturer.

SOLDERING

Anyone who installs alarm equipment should know how to solder properly. Numerous situations will exist when it will be required.

Solder, usually an alloy of lead and tin, is the material that binds metal leads on components into a circuit, both mechanically and electrically. The most commonly used solder for alarm installation work contains 60% lead and 40% tin. The abbreviation for this type of solder is simply 60/40.

Techniques

Oxides form on the surfaces of the metals being soldered. This will occur even if the metals are clean and free of oxides when you begin soldering. The oxides, and any dirt or grease that may be present, will form a barrier that can prevent a good connection. To combat the oxides, rosin flux solder is used. The flux automatically cleans away oxides. It will not clean away dirt or grease, however. Therefore 60/40 rosin-core solder is best for electronics work. Only those solders labeled safe for electronics work should be used, as some contain corrosive materials and could damage circuits.

Solder is available in a variety of sizes to meet different needs. For circuit board work, 30-gauge is the best size. A larger size may be used for soldering wire and cable.

Tools

Soldering irons or pencils are basic tools. They are available in many sizes, shapes, and wattages. Usually, a 25- to 50-watt soldering iron will be adequate. A soldering gun may be useful for working on heavy wires or cables, but it should not be used on circuit boards or electronic components.

When it is necessary to solder where no power is available, a cordless iron is handy. Powered by rechargeable batteries, it can solder up to 100 connections. If you use a cordless iron frequently, you probably need one that recharges quickly.

To keep soldering tools operating properly, they should be cleaned frequently. Use a dampened sponge or cloth to clean away dirt and oxide build-up. After the iron cools, loosen its tip about one-half turn to prevent tip seizure.

Soldering tips can be either iron-plated copper or raw copper. Raw copper tips provide better heat efficiency, but they deteriorate more quickly than iron-plated copper tips. Iron-plated tips are more expensive, but they last longer and require less maintenance.

Making Connections. Components to be soldered should be cleaned thoroughly for a good electrical and mechanical connection. After the components are cleaned, place the tip of the hot iron against the joint and hold it there for a few seconds. Sparingly feed solder between tip and joint, allowing it to flow into the connection. Then immediately remove the tip from the joint and hold the connection steady until the solder cools and solidifies. If the components are moved before the solder cools, a "cold" joint will result, giving a poor connection. (Cold connections have a dull, grainy appearance instead of a shiny, smooth one.)

When soldering electronic components such as integrated circuits and other solid-state devices, use a heat sink. Placing a heat sink, like needle-nose pliers or surgical clamps, between the component lead and the component itself will minimize heat-induced damage.

ELECTRONIC SERVICE TIPS

The following tips from the *Manual of Electronic Servicing Tests and Measurements*[1] are helpful to alarm technicians.

> Take care when using a standard VOM because it can permanently damage the small electrolytics in much of today's equipment.
>
> Dropping solid-state components, such as FETs, can damage them.
>
> As a general rule, do not exceed maximum current and voltage levels, *even temporarily.*
>
> In most cases, it's best to install the solid-state component last in the circuit. Complete all wiring and attach the DC supplies with the power switch off before making any test.

---

[1] Reprinted with permission from *Manual of Electronic Servicing Tests and Measurements* by Robert C. Genn, Jr. © 1980, Parker Publishing Co., Inc., West Nyack, NY.

When disassembling a test setup, switch off all voltages before removing the solid-state component under test.

When soldering, use long-nose pliers to hold the component leads. Hold the lead until the solder is completely cool.

Many devices, such as the FET, are greatly affected by body capacitance that can cause errors in your testing. Always use a shielded probe and keep your fingers as far away as possible from the probe tip. *Don't* use an ordinary ohmmeter test lead held in your fingers, especially when working with FETs.

Ground your soldering iron tips and use a pencil type soldering iron.

USING TOOLS

I wrote the following article, "The Well-Stocked Tool Box," which offers some suggestions for the tools needed by professional alarm installers.[2]

A common tool box is a necessary piece of equipment for the alarm installer. Believe it or not, so is a shirt with two pockets preferably with buttons and/or flaps.

The tools and supplies that an installer needs are relatively basic items, with only a few exceptions. Veteran installers will give a knowing grin as they read through this list of tools and equipment. They know the reasons for including some of the items whose value and use may not be readily apparent to a novice installer. A veteran installer knows the feeling of being in an attic and having his staple gun run out of ammunition, when the extra staples are downstairs in his tool box.

And, yes I am serious about a shirt with two pockets. The extra box of staples should have been in the left pocket. Why the buttons or flaps? If you ever had to dig through attic insulation to recover an item that fell out of a pocket you'll know.

Cases and Boxes

Before discussing specific tools and supplies, some consideration should be given to the containers in which they will be placed. There are five simple and convenient places to carry an assortment of items:

1. A tool pouch and belt
2. The pockets on a shirt
3. A tool box
4. A tool tray
5. A nearby container such as a heavy cardboard carton or orange crate

The reason that "nearby container" is included is that there are a few items that will not fit into the other four places.

A good-quality leather tool pouch with a tape thong and comfortable belt will last through years of rugged use. A large fishing tackle box makes an ideal

[2] Reprinted with permission from *Security Distributing & Marketing* magazine, February 1981. © 1981, Cahners Publishing Co.

tool box because it has numerous trays and compartments for small items. The tool tray can be used for larger, more bulky items.

Quality Pays

Purchasing quality tools saves time and money in the long run. A screwdriver with a broken tip or a drill bit that requires sharpening after each use is no bargain regardless of the purchase price. The accompanying chart [Table 12.2] shows the tools and supplies essential to routine alarm installations. Experience, and the specific types of security and fire systems to be installed, will dictate the specific types and quantities of tools and supplies required by the installer. Spare parts, such as spacers and fuses, have been omitted from the chart so we can concentrate on the primary tools and supplies.

The chart lists specific items of equipment and supplies and gives a brief description of each. It gives a recommended quantity, a recommended location for storage and an estimated cost.

The quantities and locations are based on personal experience and personal preference. The key point is that all of the frequently used tools and supplies should be easily accessible to the installer to avoid unnecessary delays.

The equipment costs listed here are estimates based, for the most part, on local sources in Oklahoma City. Prices will vary from city to city and from supplier to supplier. They should be accurate enough, however, to give an installation supervisor or manager an approximate idea of the capital outlay required to outfit an installation vehicle or crew or an individual installer.

What's the Use?

The uses for most of the items listed in the chart are obvious. Some of the items, however, may require an explanation.

A scratch awl is a handy tool for punching holes in sheetrock or panelling, starting screws in wood, and scoring wood, metal, or plastic surfaces. A short piece of stiff wire (coat hanger) is valuable for pulling wire through hollow walls instead of using a long snake (fish tape).

Disposable brushes are useful for cleaning terminals and contacts as well as for applying varnish to foil. It is much simpler, and less expensive, to discard an inexpensive brush than to spend valuable man-hours cleaning an expensive one.

Selection of a wire cutter/stripper is a matter of personal preference. Personally, I like a small pair that adjusts, with a nut, to fit different gauges of wire. Once the cutter is set to cut 22-gauge wire, larger sizes can be cut with a little practice.

Sometimes a flashlight is easier to handle in an attic or above a suspended ceiling than a larger lantern. A leather holster on your belt insures that light is always within reach. The police-type flashlight holster can be purchased at stores that sell supplies for law enforcement and security officers.

Book labels, the type used on the spines of the books at the library, can be purchased at most office supply stores. Using them to mark wires at several places

Table 12.2 Typical Installation Tools

Item	*Recom-mended Quantity*	*Recom-mended Location*	*Esti-mated Cost ($)*
Adapter, U-blade, for 3-wire cords to 2-wire receptacles	2	TB	1.00 ea
Anchor, plastic, hollow-wall, small	100	TB	2.75/C
Anchor, plastic, hollow-wall, large	100	TB	3.25/C
Awl, scratch, 3½″ blade	1	TP	4.00 ea
Bit, drill, long, ¼ × 12″	1	TB	4.00 ea
Bit, drill, long, ¼ × 30″	1	NB	6.00 ea
Bit, drill, masonry, ¼ × 4″	2	TB	1.50 ea
Bit, drill, masonry, ⅜ × 4″	2	TB	2.00 ea
Bit, drill, set, high-speed, jobber length, 1⁄16 to ⅜″	1*	TB	15.00/set
Bit, drill, set, paddle, ½ to 1″	1	TB	10.00/set
Blade, razor, single-edge	5	TB	0.20 ea
Brush, disposable, for applying varnish	5	TB	0.18 ea
Cable, test ("jumper"), alligator-clip ends	10	TB	0.50 ea
Chisel, wood, set (1 each: ¼″, ½″, ¾″, 1″)	1	TT	15.00/set
Clamp, wire/cable, nylon, ⅛″ capacity	100	TB	2.50/C
Clamp, wire/cable, nylon, ¼″ capacity	100	TB	3.00/C
Clip, cable (Hiatt clip), 0.14″ capacity	100	TB	2.80/C
Clip, cable (Hiatt clip), 0.20″ capacity	100	TB	3.25/C
Clip, cable (Hiatt clip), 0.31″ capacity	50	TB	4.50/C
Cord, electrical, extension, 3-wire, 15′	1	NB	7.00 ea
Cord, electrical, extension, 3-wire, 25′	1	NB	18.00 ea
Cord, electrical, extension, 3-wire, 50′	1	NB	28.00 ea
Cutter, end, 4½″ long, ⅞″ cut	1	TT	12.00 ea
Cutter/stripper, wire, small, adjustable	1	TP	2.50 ea
Drill, electric, ⅜″	1	TT	25.00 ea
File, flat, bastard, 8″	1	TB	2.50 ea
File, round, bastard, 8″	1	TB	2.50 ea
File, triangle, bastard, 8″	1	TB	4.00 ea
Flashlight, 2-cell, with belt holster	1	TP	10.00 ea
Hammer, ball-peen, 12-ounce	1	TT	4.75 ea
Hammer, electrician's, 18-ounce	1	TT	9.00 ea
Hanger, coat, wire (straightened)	1	TT	—
Iron, soldering, electric, 25- 40-watt	1	TT	10.00 ea
extra tips for above	2	TB	1.50 ea
Knife, fixed blade (hunting), with sheath, 3–4″ blade	1	TP	11.00 ea
Labels, book, white, paper, ¾ × 1″, 20 per sheet	10	TB/SP	3.00/bx
Lantern, battery-powered	1	TT	12.00 ea
Meter, volt-ohm, pocket-size, with test leads	1	SP	10.00 ea
Molding (Dekduct), self-adhesive, gray vinyl wire duct ⅛″ capacity, 4′ long	10	NB	1.80 ea
Molding, as above, ¼″ capacity, 4′ long	5	NB	2.50 ea
Nails, assorted sizes	100	TB	2.75/C
Nutdriver, set (small), 3⁄32″ to ⅜″	1	TB	11.00/set
Pen/pencil	1*	SP	0.15 ea
Plier, electronic, diagonal cutter, 5″	1	TB	6.00 ea
Plier, long-nose, with cutter, 5″	1	TP	6.50 ea
Plier, slip-joint, 4″	1	TP	3.00 ea
Plier, slip-joint, 10″	1	TB	5.50 ea
Plier, tongue-and-grove (pump), 10″	1	TB	7.00 ea

Table 12.2 Typical Installation Tools (*continued*)

Item	*Recommended Quantity*	*Recommended Location*	*Estimated Cost ($)*
Rule, tape, retractable, with belt clip	1	TP	5.50 ea
Saw, hack	1	TT	12.00 ea
Saw, keyhole	1	TT	9.00 ea
Screwdriver, Phillips, #0 point, 3″ blade	1	TP	2.00 ea
Screwdriver, Phillips, #2 point, 4″ blade	1	TB	3.00 ea
Screwdriver, Phillips, #4 point, 8″ blade	1	TB	5.00 ea
Screwdriver, standard, with pocket clip, 1/8 × 3″	1	SP	1.25 ea
Screwdriver, standard, 3/16 × 4″	1	TP	2.00 ea
Screwdriver, standard, 3/16 × 10″	1	TP	3.25 ea
Screwdriver, standard, 1/4 × 8″	1	TB	3.00 ea
Screwdriver, standard, 3/8 × 10″	1	TB	5.00 ea
Screwdriver, stubby, standard, 1/4 × 1 1/4″	1	TB	2.25 ea
Screw, machine, with nuts, 8/32 × 1/2″	25	TB	3.50/C
Screw, machine, with nuts, 8/32 × 1″	25	TB	4.00/C
Screw, sheet metal, #6 × 1/2″	50	TB	1.50/C
Screw, sheet metal, #6 × 3/4″	100*	TB	1.75/C
Screw, sheet metal, #6 × 1″	100	TB	2.00/C
Screw, sheet metal, #6 × 1 1/2″	50	TB	3.25/C
Screw, sheet metal, #8 × 3/4″	100*	TB	1.75/C
Screw, sheet metal, #8 × 1″	100	TB	2.00/C
Screw, sheet metal, #10 × 1″	50	TB	2.50/C
Screw, wood, assorted sizes	50	TB	3.00/C
Silicone rubber, clear, 3-ounce tube	1*	TB	3.00 ea
Snake, wire (fish tape), assorted lengths	5	NB	0.35/ft
Solder, rosin core, 1-pound roll	1	TT	11.00/rl
Staples, wire, 1/4 × 3/8″, galvanized, box of 1,000	1*	TT	1.50/bx
Staples, wire, 1/4 × 3/8″, beige, box of 1,000	1*	TT	1.50/bx
Staples, wire, 1/4 × 7/16″, galvanized, box of 1,000	1*	SP	1.75/bx
Staples, wire, 1/4 × 7/16″, beige, box of 1,000	1*	TT	1.75/bx
Tacker, wiring (staple gun), uses 1/4″ crown staples	1	TP	25.00 ea
Tape, electrical, plastic, roll	1*	TP	0.50/rl
Ties, cable, self-locking, 4″	50	TB	4.50/C
Ties, cable, self-locking, 7″	50	TB	6.00/C
Varnish, spar, clear, 4-ounce can	1	TT	1.50 ea
Wire, 2 conductor, 18-gauge stranded, clear plastic, 500′ roll	1*	NB	28.00/rl
Wire, 2 conductor, 20-gauge stranded, clear plastic, 500′ roll	1*	NB	19.00/rl
Wire, 2 conductor, 22-gauge stranded, clear plastic, 1000′ roll	1*	NB	19.00/rl
Wire, 4 conductor, 22-gauge solid, jacketed, 500′ roll	1	NB	25.00/rl
Wrench, adjustable, 4″, 1/2″ capacity	1	TP	5.50 ea
Wrench, adjustable, 8″, 15/16″ capacity	1	TB	6.50 ea
Wrench, adjustable, 12″, 1 5/16″ capacity	1	TT	15.00 ea
Wrenches, hex key, set with fold-up handle	1	TB	3.00/set

* Frequently used items, extras should be readily available.

Abbreviations: C (hundred), NB (nearby), TB (tool box), TP (tool pouch/belt), TT (tool tray), SP (shirt pocket).

along the wire run will reduce the time necessary to identify them during troubleshooting. A sheet of twenty labels should be carried in your shirt pocket.

The volt-ohmmeter is another item where personal preference plays an important role. My personal preference is a small pocket-size unit. Large meters have a place on the workbench; small ones, in the field.

The assortment of screwdrivers may seem large, but each of them has a place in installation work. A small screwdriver that clips in a shirt pocket is even more useful if it has a 110-volt AC tester built into the handle.

Dekduct™ looks like quarter-round molding. It is hollow, allowing wires to be concealed inside. It can be painted so that it will blend in with the surroundings. It is a preferable alternative to having totally exposed wiring.

Silicone rubber comes in a three-ounce tube and in a large tube for caulking guns. Its uses are many: sealing cracks and holes, attaching circuit boards (such as siren drivers) to control boxes, adding extra insulation to terminals and wires, and gluing a variety of items together. A small amount of silicone rubber on the self-adhesive pad of a contact switch that has lost its adhesiveness will bond the switch to almost any surface.

ADDITIONAL HANDY TOOLS

Since I wrote that article, I have found several other useful items that I would consider valuable additions to an installer's complement of tools and equipment.

Long, up to six feet, flexible drill bits, such as those manufactured by Diversified Manufacturing & Marketing Co. and Alarm Products of California, Inc., simplify drilling and wire-pulling tasks. They are available in a variety of lengths and diameters to meet the needs of most alarm installers.

Also good for pulling wire is an under-carpet fish tape manufactured by Alarm Products of California, Inc. Running wire under carpet usually is not recommended, especially in high-traffic areas, because it might become damaged. Placing it under a carpet near a baseboard where foot traffic is unlikely, however, is made much simpler with the under-carpet tape. Moreover, the tape's flat design makes it ideal for pulling wire inside walls where sheet insulation is present or for running it above a suspended ceiling.

Another good wire "fishing" device is ComTec Industries' Fish Grabber, a five-foot-long tool made with flexible, interlocked armor casing. The plastic-encased wire loops expand to sixteen inches in diameter once inside a hollow wall, providing a large target area to grab the wire being fished. The device only requires a ⅜-inch hole for insertion into the hollow wall. Figure 12.4 shows the Fish Grabber in its normal and expanded positions.

A handy time- and labor-saving device for applying foil tape to windows is Foilmaster's foiling tool (Figure 12.5). With some practice, you can foil a window in minutes. The device's movable arm lets you select the appropriate distance from the edge of the window to apply the foil, then helps ensure that the foil is applied in a straight line. Also, an optional template set allows you to apply foil patterns with rounded corners.

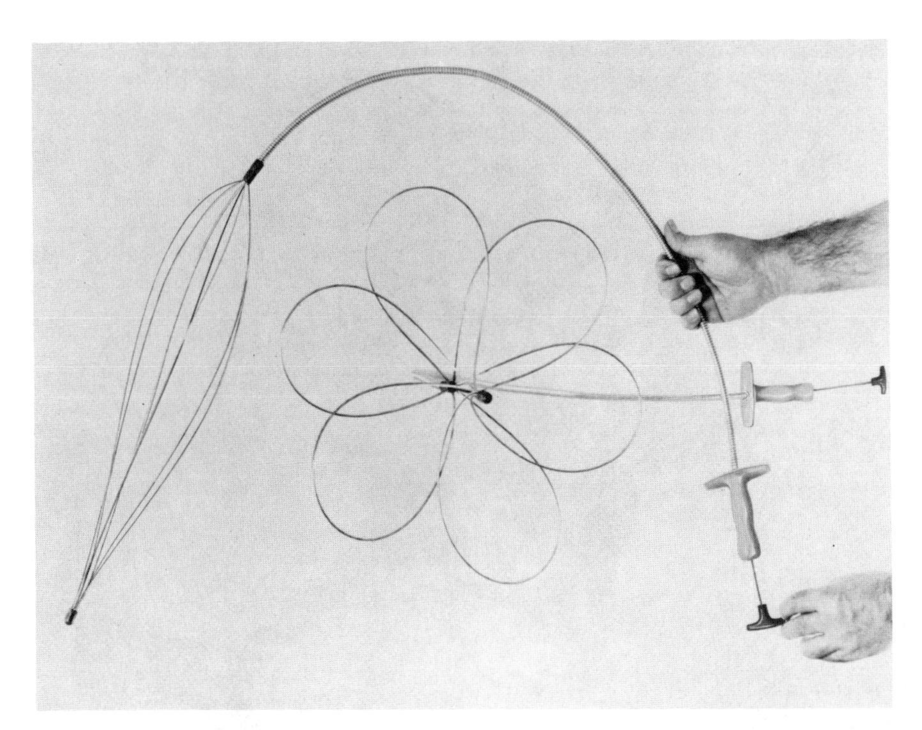

Figure 12.4 The Fish Grabber simplifies placing wires inside hollow walls. Once inserted, the device expands so that wires or fish tapes can be snagged easily. (Photo courtesy of ComTec Industries, Inc.)

To save even more time and labor, consider a cordless drill. Skil Corporation manufactures several models, so you can select one that will meet your specific requirements. One model is a high-performance ⅜-inch cordless drill/screwdriver that is powered by rechargeable batteries. A removable quick charge power pack can be recharged from zero to a full charge in one hour. With a second power pack as an accessory, downtime for recharging is virtually eliminated. Skil products are available at most hardware stores.

To help ensure your safety, a ground fault interrupter (GFI) is highly recommended. A portable model can be attached to your belt. The purpose of a GFI is to shut down a power tool before a dangerous electrical shock can occur.

ALARM SYSTEM COMPONENTS

In addition to having a basic understanding of a few common electronic components, tools, and equipment, an understanding of system components is necessary. Alarm systems consist of three general categories of components: detectors, processors, and annunciators. Each of these components is discussed in more detail in the following chapters.

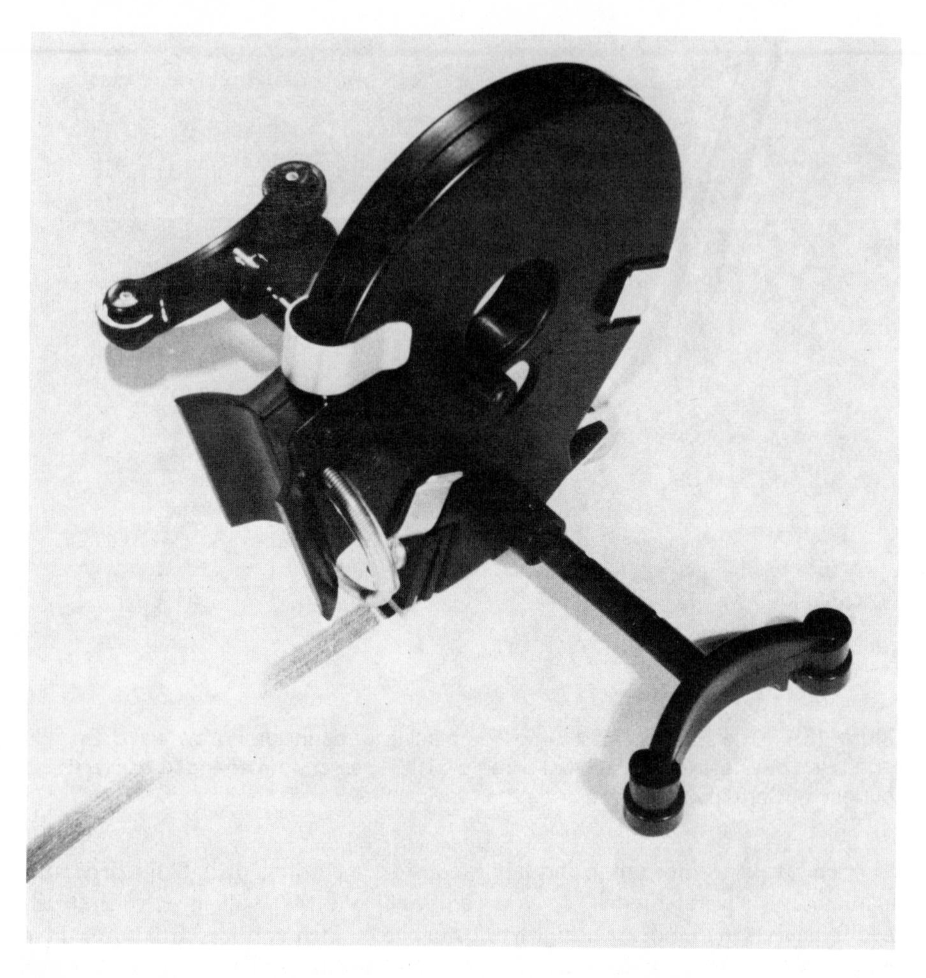

Figure 12.5 This foiling tool can greatly speed up foil application time. It applies foil tape quickly and bonds it more firmly than traditional methods. (Photo courtesy of Foilmaster, Inc.)

A detector does just what its name implies—it detects something. For example, a magnetic contact switch can detect a door opening and a passive infrared (PIR) detector responds to changes in ambient temperature.

Once a detector has responded to a stimulus, it signals a processor, or control panel. The processor analyzes the signal and determines if it is valid or invalid. If the signal is valid, the processor sets off an annunciator, like a bell, siren, or light. It also can trigger a device, like a tape dialer or digital communicator, which would announce the system's status at a remote location by telephone lines.

Some alarm system devices contain several components. For example, a siren or digital communicator could be built into the control panel. The control panel/

Figure 12.6 Fyrnetics' 8-channel microprocessor receiver combines control functions with a digital communicator as an annunciator. (Photo courtesy of Fyrnetics, Inc.)

annunciator combination shown in Figure 12.6 has both a siren and digital communicator built into the unit. Also, a detector (an ultrasonic motion detector, for example), processor, and annunciator could be contained in one package, such as the one shown in Figure 12.7.

Many of the newer control panels are equipped with built-in digital communicators. As more alarm equipment becomes microprocessor-based, it is likely that we will see more control panel/digital communicator combinations.

Figure 12.7 Master Lock's Ultrason-II is a combination unit that includes an ultrasonic detector, control, and horn. (Photo courtesy of Master Lock Co.)

CONNECTING THE COMPONENTS

In a literal sense, components can be connected by wires. In a more figurative sense, they can be connected by radio frequency (RF) or ultrasound, that is, wireless. To date, however, a totally wireless system does not exist. Wires are needed to connect power and some annunciation devices, like sirens or bells, to the control panel.

Wired Systems

Most burglar alarm protective circuits use series-wired, normally closed (NC) detection devices. Many control panels also accept parallel-wired, normally open (NO) devices; few accept only normally open circuits. Many of the newer panels have combination normally closed/normally open circuits requiring an end-of-line (EOL) resistor for line supervision.

In the NC circuit, often referred to as just closed circuit, the momentary opening of the protective device's switch or relay activates the alarm, if the system is armed. In the NO (open circuit) protective circuit, the reverse is true: a contact or relay closure trips the system. Figure 12.8 shows simple NC and NO protective circuits.

Typically, fire alarm systems use NO circuits. Burglar alarm systems may use only NC or both NC and NO circuits.

The circuits shown in Figure 12.8 have drawbacks. If the NC circuit is shorted at some point along the protective loop, all devices beyond that point

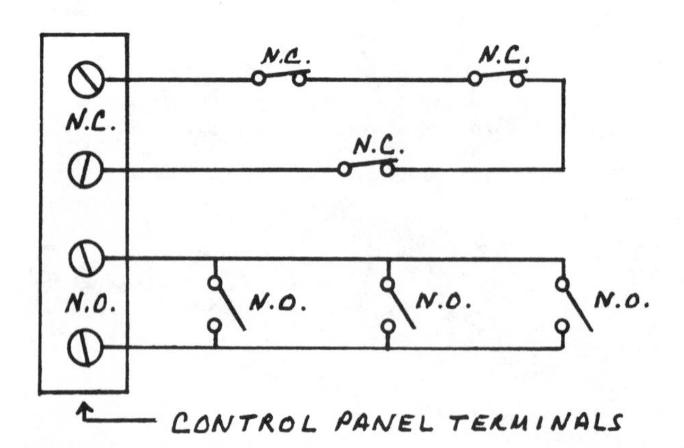

Figure 12.8 Simple, normally closed (NC) and normally open (NO) protective circuits or loops. Opening the closed circuit or closing (shorting) the open circuit will trigger an alarm if the system is armed.

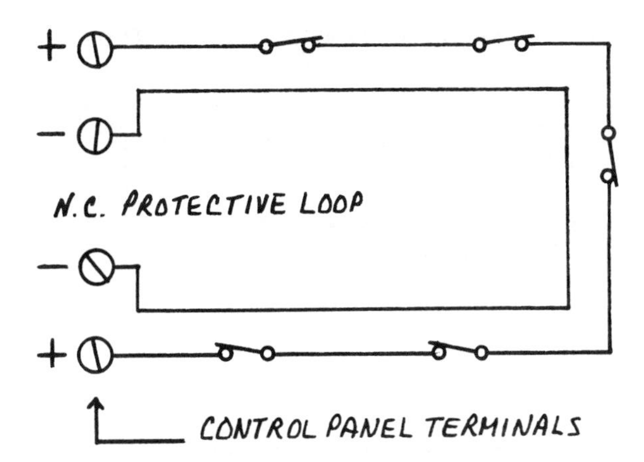

Figure 12.9 Some panels are designed for NC return loops. Detection devices should be placed on the positive side of the protective circuit.

will be rendered ineffective. Similarly, in the NO circuit, an open or broken circuit will defeat part of the loop.

To overcome this drawback in the NC circuit, some panels are built for a two-wire (return circuit) loop. Figure 12.9 shows a simple return circuit. Note that detection devices should be placed on the positive side, or "leg," of the circuit. Polarity is important.

End-of-line resistor circuits provide line supervision for NC, NO, and combination protective loops. Figure 12.10 shows simple EOL resistor circuits. Essentially, the circuit ends at a resistor with a specified resistance value. An increase or decrease (short or open) in resistance will trigger an alarm.

Wire and Cable. Insulated, single-conductor wire is the most common type of wire, but it has few applications in security systems. Two-conductor insulated wire, either parallel or twisted pair, is more common. Multiconductor wire is popular with many alarm dealers, with jacketed, four-conductor wire (sometimes called telephone wire) being widely used. Specialized applications may require special types of wire.

Twisted-pair wire is recommended for protective loops because the twisting provides some shielding from outside interference. Two-conductor parallel wire, often called "zip" cord, can be used for short power runs and to connect bells or sirens to the control panel.

Requirements for burglar alarm wire are set forth in the *National Electrical Code* (*NEC*). Fire alarm wire must meet numerous requirements, however. It is recommended that you study the *NEC* as well as state and local building codes and fire ordinances before installing fire alarm systems.

Unless the *NEC* or local codes specify otherwise, you may use either stranded or solid wire. Using one or the other will depend on the particular application and your personal preference.

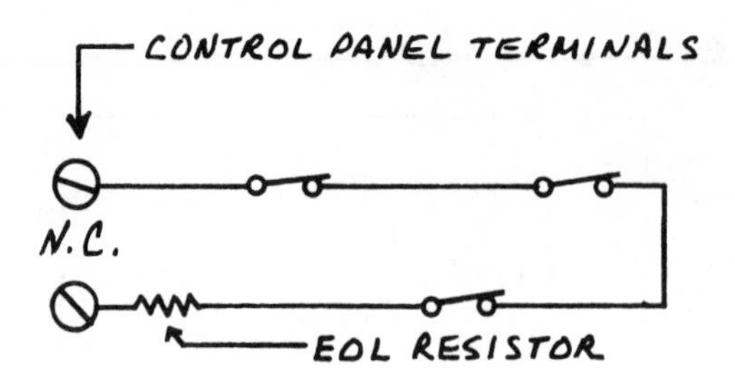

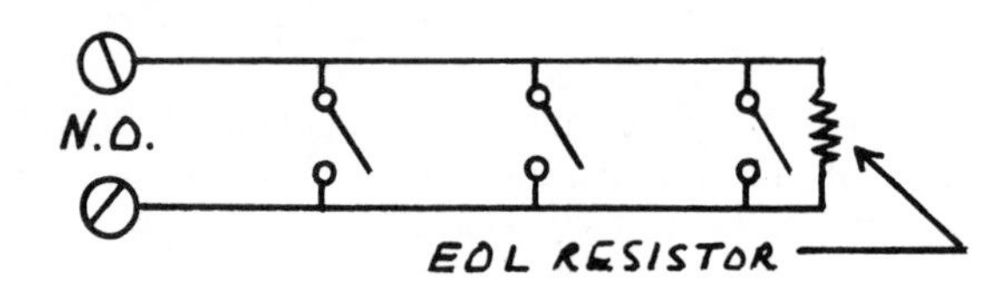

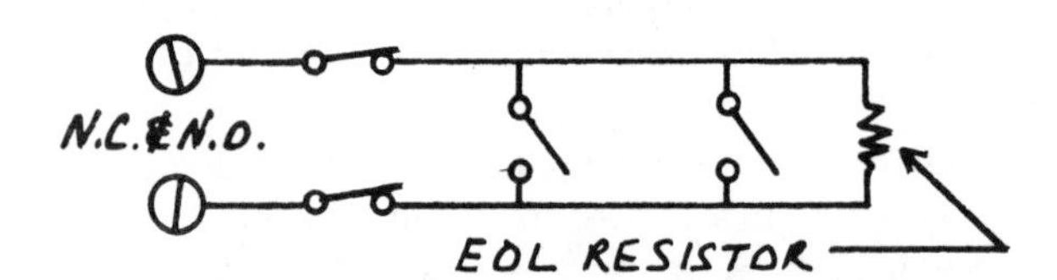

Figure 12.10 End-of-line resistor circuits afford better line supervision. If the protective circuit's resistance goes above or below certain specified limits, the alarm will sound. Attempts to cut or short the circuit are made more difficult because of the end-of-line resistor.

Wire Resistance. Whichever type of wire you select, be sure that it is large enough to minimize resistance problems. Using the correct wire size is important to efficient alarm system operation. If in doubt, select one that might be too large for the job instead of one that might be too small; it may be more expensive, but you will have fewer problems later.

The larger the wire, the smaller its gauge number. Conversely, the smaller the wire, the larger its gauge number. The smaller the wire, the greater its resistance. Table 12.3 shows resistance values for wire gauges commonly used in alarm installations. Keep in mind that you will have to double the resistance values if you are using a two-conductor wire in the protective circuit.

Table 12.3 Resistance of Copper Wire

*Gauge (AWG)**	*Resistance (Ohms) per Foot*
24	0.026
22	0.016
20	0.010
18	0.006
16	0.004

* American Wire Gauge

The resistance values shown are per foot of single-conductor copper wire. To estimate the resistance in a wire run, select the appropriate value for the gauge of wire you are using. Then multiply the length of the wire run (feet) by the resistance value. If you are using two-conductor wire, you will have to double the resistance value. For example, a wire run of 100 feet of two-conductor wire is the same as a 200-foot run of single-conductor wire. If you are using 22-gauge wire, the calculations would be as follows: $0.016 \times 100 = 1.6$ ohms $\times 2 = 3.2$ ohms total resistance.

Wire that is too small for the job will create too much resistance and result in voltage loss. In some cases, when the wire is much too small, it will become warm or hot. Not only may a fire hazard be created, the lower voltage may cause the system to malfunction.

Wireless Systems

Recent technologic advances have made wireless alarms stiff competitors with their hard-wired counterparts. Wireless alarms have been available for years, so the concept of using a radio frequency (RF) to transmit alarm signals from a detector to the control panel is not new. What is new are high-tech—supervised—wireless systems.

The older, nonsupervised wireless systems had two drawbacks. First, you had to check each transmitter's battery periodically to be sure that it had enough power. Second, you could not determine circuit status; that is, you could arm the system and have one or more doors or windows open and not know it.

Supervised wireless alarms solve those problems and offer several other benefits. In addition to checking battery condition and circuit status, they report tampering and transmitter status. In some systems, each transmitter reports its condition periodically. Failure to report initiates a trouble signal at the control panel or central monitoring facility.

The most clear-cut benefit of wireless alarm systems is labor savings. You can install them faster, so you can install more of them. Therefore you can generate more income—in two ways.

First, because you can install more systems, you can earn more in installation charges. The prices of hard-wired versus wireless systems are comparable. With a hard-wired system, you would have lower equipment costs and higher labor costs. The reverse is true with wireless systems. Some dealers claim that they can install four wireless systems in the same time that it takes to install one hard-wired system. In that case, if you installed one hard-wired system per week for $2,000, you would have the capability of installing four wireless systems, increasing your weekly sales to $8,000.

The second way to increase revenue with wireless alarms is through your lease service and/or monitoring fees. Using our four-to-one ratio, if you were installing one system and generating $30 per month in recurring revenue, you could increase that to $120 per month by installing four wireless systems.

Wireless alarms are coming of age and it is definitely worth your time and effort to evaluate the potential they offer. You may discover that you will benefit by switching from hard-wired to wireless systems, or at least by incorporating some wireless devices into your hard-wired systems.

Control Devices 13

The heart or, more appropriately, the electronic brain, of an alarm system is its control panel. The panel can be simple or complex; it can perform few functions or many; and it can be adequate or inadequate for the job you want it to do.

Choosing a control panel that is appropriate for a particular installation is an important task. To make a wise, cost-effective choice, you need to know what you want the panel to do. A panel with too few features and functions may not meet your customer's needs or it could cause future service problems. Too many features may confuse your customer, require additional installation time, or be too costly.

Control panels for residential and commercial alarm systems are available in a wide variety of sizes and shapes and with myriad functions. Some are combination panels, including fire alarm circuits in addition to intrusion alarm circuits. Others are stand-alone systems for either intrusion or fire alarms.

Stand-alone, commercial fire alarms can be complex, requiring a high degree of technical and installation expertise. Many of their functions and features are similar to those of intrusion or combination intrusion/fire alarm systems, however. In this chapter, our discussion primarily includes intrusion and combination alarm system control panels.

SELECTING FEATURES

The job you want the panel to do should dictate which panel you use. Theoretically, if you custom-design security systems to meet the needs of your customers, you might use many control panels, selecting the most appropriate one for a particular application. This usually is not practical, however. The more types of panels you use, the less familiar your installation, service, and troubleshooting crews can become with the equipment. Using fewer types of panels also reduces your inventory requirements.

Usually, dealers can select two or three types of control panels that will meet 90% of their installation needs. Only rarely—the other 10% of the time—will they have to use different ones.

Whether or not your two or three standard panels are manufactured by the same company is not as important as the features they offer. That their features meet your customers' security needs and your installation, operation, and service requirements is what is important. When selecting your standard panels, consider the features most often needed for the types of systems you install. For example, if your community has a noise-abatement ordinance, you will need panels with an automatic cut-off feature to silence the bell or siren. If your usual procedure is to divide an alarm system into zones, your standard panels should accommodate your zoning needs.

Automatic cut-off, reset, and rearm features are useful and are included on many control panels. Intrusion or combination intrusion/fire alarm panels often include entry and delay circuits, instant circuits, a twenty-four-hour panic circuit, a fire circuit, an auxiliary power output for protective devices, a charging circuit for rechargeable batteries, dry (relay) and voltage alarm outputs, and remote arming capability. Of course, the panel should be UL listed.

Protective Zones

The panel you select may or may not have several zones. If it does not, you can always add a zone annunciator to the alarm system. (Zone annunciators are discussed in more detail in Chapter 15.)

When selecting a multizone control panel, exercise caution. Some manufacturers claim that their panels are multizone models when in reality they are counting protective circuits as zones. With this way of thinking, for example, a panel with an entry/exit delay circuit, two instant circuits, a twenty-four-hour panic circuit, and a fire circuit could be called a five-zone panel. In my opinion, a true multizone control panel provides for individual zone annunciation and control, and the example does not meet those criteria.

Zone identification is important. It helps you identify the area in which the alarm occurred or the specific detector that caused the alarm. It helps pinpoint trouble spots, too, so you can provide efficient service.

Supervised Circuits

An increasingly popular feature is circuit supervision. Not only does a supervised circuit make the system more difficult to compromise, it can alert your customer or your central station to a problem before the system is armed.

Supervised protection circuits use resistors to effect supervision. Any change in circuit resistance, whether an increase or a decrease, beyond a certain resistance level will cause an alarm or trouble signal to be annunciated. Most supervised circuits allow the use of both normally closed and normally open detection devices. Thus either opening (breaking) or closing (shorting) the circuits will trigger an alarm if the system is armed (see Figure 12.10 for a diagram of simple supervised protective circuits).

Time-Delay and Instant Circuits

At least one of the control panel's intrusion circuits should have an adjustable time delay. My preference for both commercial and residential alarms is to use a control panel with an entry/exit delay. One or more doors can be included in the delay circuit, allowing the homeowner or businessperson to enter the premises and disarm the system only through these doors; entry through any other protected point or area triggers an immediate alarm.

Using a control panel with an entry/exit delay means that an exterior control device such as a key switch or digital keypad is not needed, which lowers the risk of tampering. Of course, if your customer prefers an exterior control station or if his insurance carrier requires one, the entry/exit delay feature should not be used.

Most of the time, a thirty-second entry delay and a sixty-second exit delay are adequate. Upon entry, the alarm user should be able to reach the control panel or control station and disarm the system without being hurried. Determine how long it takes to walk from the entry door to the control point to establish the entry delay time. It may be a good idea to add five or ten seconds to that time as a margin for error. Only in rare instances should the entry time exceed sixty seconds.

In a sense, exit time is not as critical as the entry time delay. Here, too, you should determine the time required to arm the system, walk to the exit, and close and lock the door. Adding an extra ten to twenty seconds will provide plenty of time for your customer to pick up a package, briefcase, or purse after arming the system and still exit before the time delay expires. Sixty seconds are usually adequate and exit time should rarely exceed ninety seconds.

Except for the protective devices used on entry/exit doors or space-protection devices located in areas that must be crossed to reach the control device, all other protective devices should be on an "instant" circuit. That is, the alarm will sound instantly if the circuit is violated.

Other Protective Circuits

In addition to entry/exit delay and instant intrusion protective circuits, two other protective circuits should be included in your standard combination panel: twenty-four-hour panic and fire circuits. Both are useful in residential and commercial installations.

Panic or Hold-up Circuits. In a residential application, the twenty-four-hour circuit is usually labeled panic or emergency; in a commercial system it is called a hold-up circuit. Regardless of what it is called, it should be a twenty-four-hour circuit. It should be operable whether the alarm system is armed or not.

Panic/hold-up circuits usually only accept normally open devices. Whether the circuit triggers a loud or silent alarm depends on the control panel's design and your customer's security needs. Both have applications in residential systems. In a commercial system, however, it should be a silent hold-up alarm to prevent startling a robber. (More information about the components of hold-up alarms is provided in Chapter 16.)

Fire Protection Circuits. Unless you usually install stand-alone, commercial-grade fire alarm systems, you should consider selecting a combination control panel that includes at least one fire circuit. Adding this to a security system provides your customer with a more complete package of protection.

The fire circuit should be supervised and have a separate alarm output. It also should be a twenty-four-hour circuit so it cannot be turned off accidentally

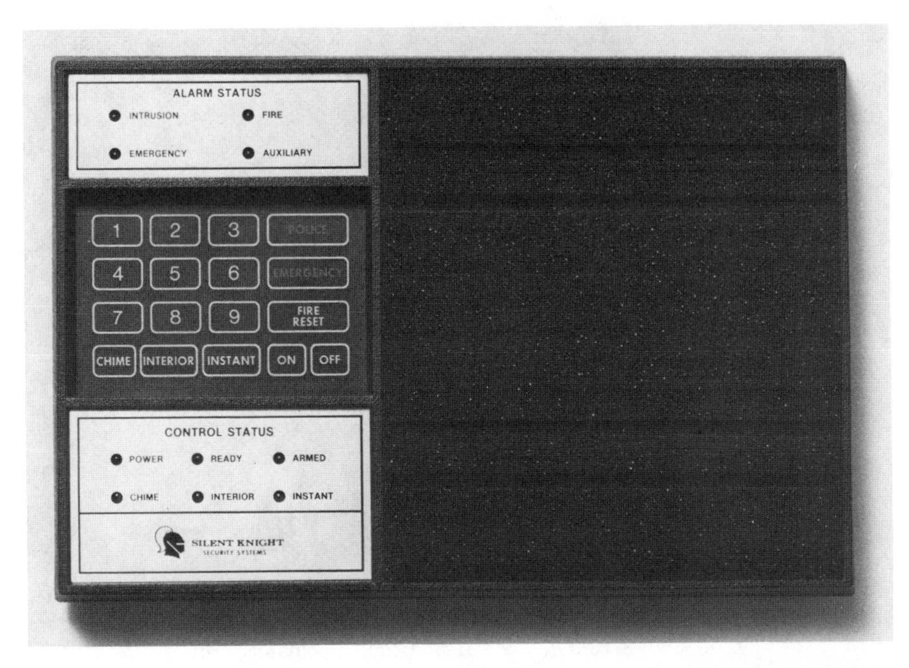

Figure 13.1 Control panels that have visible and/or audible status indicators help pinpoint alarms and trouble spots. (Photo courtesy of Silent Knight Security Systems.)

by the alarm user. The combination control panel shown in Figure 13.1 has a light-emitting diode (LED) to annunciate alarm status; in an alarm condition its siren provides distinctive sounds to indicate the type of emergency.

Fire alarm circuits should have an audible and/or visual signal to indicate trouble with the fire circuit. It also should have a switch or button to reset smoke detectors or other fire detection devices manually.

Prealarm Warning

To remind your customer to disarm his system and to reduce inadvertent alarms, a prealarm output is helpful. Used with the entry/exit delay circuit, a prealarm warning sounds immediately upon entry. It should continue to sound until the entry time has expired and the system goes into an alarm condition.

The prealarm warning serves two purposes. First, it reminds the alarm user that he must disarm his system. Second, it notifies an intruder that something is happening or is about to happen, and if he stays until the entry delay time has expired, he will be greeted by a siren or bell.

A beep, buzz, or musical tune can be used as a prealarm warning. I prefer beepers and buzzers for commercial installations and musical tunes for residential

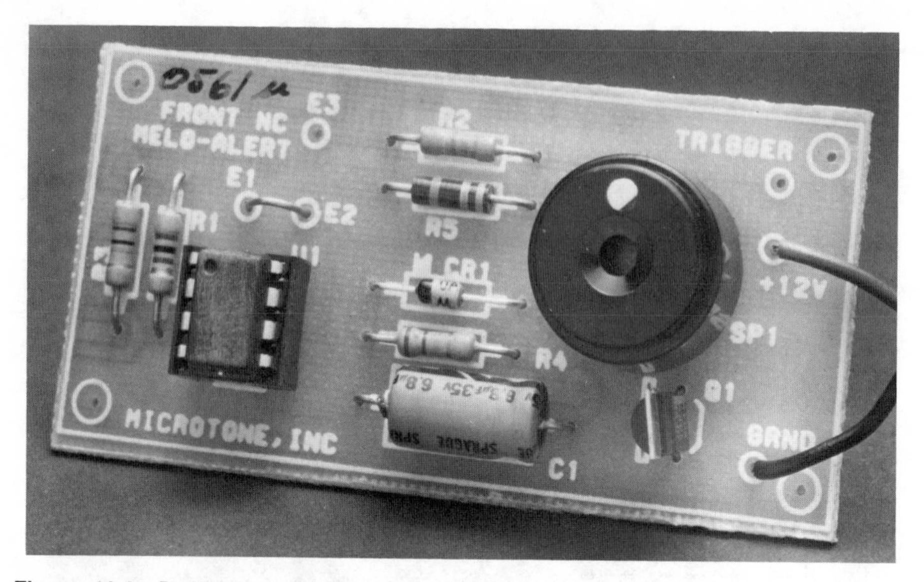

Figure 13.2 Prealarm warning devices can emit beeps, buzzes, or musical tones. When connected to a control panel's entry warning voltage output, this small circuit board plays "Home Sweet Home." The device can be installed inside the control's cabinet or flush-mounted in a single-gang electrical box with a specially designed plate. (Photo courtesy of Microtone, Inc.)

applications; when the homeowner enters, he is greeted by the sounds of "Home Sweet Home" (see Figure 13.2).

In addition to providing voltage outputs for audible prealarm devices, some control panels have visual prealarm annunciation. For example, the panel's loop status LED may start flashing upon entry. Visual annunciation is not as noticeable as audible annunciation, however.

Automatic Cut-off and Reset

Most newer control panels have provisions for automatically shutting off the alarm and resetting the system after a predetermined time. This is especially important for loud alarms in communities with noise abatement laws.

Systems that must be manually reset can be a nuisance. I have heard numerous stories about alarms that sounded all day, all night, or all weekend until the alarm user returned to disarm and reset the system.

Some panels have factory-set cut-off times. Others have adjustable controls that allow the dealer/installer to select the appropriate time for the system to shut down and reset. I prefer the latter. If an alarm system is monitored, the siren or bell only needs to sound for a few minutes to notify neighbors of a problem. The central station operator will dispatch the police or fire units upon receipt of an alarm signal and then notify the alarm user. With an adjustable cut-off timer, you can select the appropriate time for the alarm to sound: a few minutes for a residential system, longer for a commercial system.

Alarm Outputs

Alarm output features vary. Some panels provide relay contacts and some provide a voltage output; many provide both. Some provide separate outputs for intrusion, panic/hold-up, and fire alarms. Many of the multizone units provide individual zone outputs; some have separate outputs for digital communicators. The types of systems you install will determine which alarm outputs best suit your needs.

Voltage Requirements

Intrusion alarms usually operate on 6 or 12 volts; the latter is more common with newer control panels. Most fire alarm systems require 24 volts.

Although no industry standard currently exists, 12-volt systems are recommended. The majority of detection devices, like ultrasonic and passive infrared detectors, require 12 volts to operate.

It is extremely important to read and follow the manufacturer's instructions concerning control panels' system power requirements. Using the wrong power supply can cause system failure or equipment damage.

Operating Voltage. Most control panels require a 12-VAC, 20- or 40-VA (volt-amps), class II, UL listed transformer for primary power. Check the control panel's specification sheet to be sure that the transformer you are using is suitable. Input power cables or wires should be as short and straight as possible, and be 18-gauge or larger. Long wire runs result in voltage drops and excessively long runs may cause the voltage to be low enough to affect the panel's operation.

Auxiliary Power Output. Power should be available from the panel to operate protective devices like smoke detectors and digital communicators. The auxiliary output voltage should be at least 600 milliamps and care should be taken not to overload the circuit. If more power is needed than can be supplied from the auxiliary output, use a separate power supply.

Standby Power. Secondary, or standby, power is needed in case primary power fails. The control panel should automatically switch to standby batteries in case of an AC power failure.

Standby power can be dry cell or rechargeable batteries. Dry cell batteries must be checked and replaced periodically; rechargeable batteries last many times longer and do not require such frequent checking and maintenance.

The control panel you select should have a built-in charging circuit to maintain its rechargeable battery at a full charge at all times. Be sure that the rechargeable

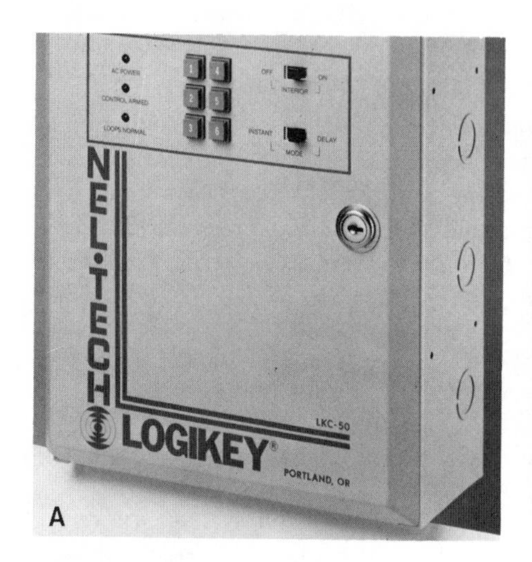

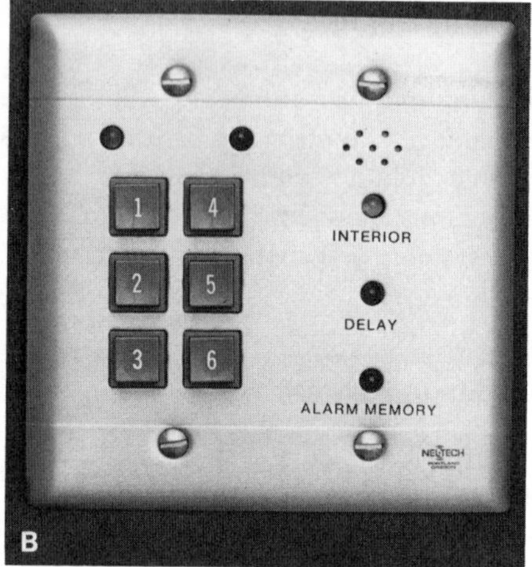

Figure 13.3 **(A)** Some control panels have built-in keypads. An important feature to consider is a control panel's remote arming capability. (Photo courtesy of Nel-Tech Development, Inc.) **(B)** Other control panels are designed to be operated entirely from remote locations. This Nel-Tech LKC-60 is controlled from keypads. (Photo courtesy of Nel-Tech Development, Inc.)

battery you select is compatible with the control panel, otherwise, either or both may be damaged.

Remote Control

A very useful control panel feature is its remote arming capability. If the system must be armed and disarmed from various locations, remote key stations or digital keypads are a must. Without this capability, the system must be operated from the control panel.

Some control panels such as Nel-Tech's LKC-50 (Figure 13.3A) can be operated from the main panel or from a remote keypad. If the control panel is to be installed in a location that is convenient for the customer, remote control points may not be needed, then a control panel with a built-in key switch or keypad will suffice.

If the control panel is to be installed in a closet or other out-of-the-way location, a system operated entirely by remote control devices will be your best choice. For example, Nel-Tech's LKC-60 system has no user-operable controls on its cabinet. Instead, all system functions are controlled remotely by digital keypads such as the one shown in Figure 13.3B.

Fail-Safe Arming

Regardless of where the system is armed and disarmed, the control panel should include a fail-safe, or arm-inhibit, feature. Unless all protective loops are normal, the alarm user cannot arm the system. This feature helps prevent false alarms by requiring that all protective loops be checked and properly set before the system can be armed.

Appearance

If a control panel is to be installed in a storage room or closet, its appearance is of little importance. It need be nothing more than a plain metal cabinet with a lock. If it will be visible to a homeowner and his family or to a businessperson and his employees or customers, appearance is important. The unit should be attractive and blend well with the surroundings.

HARD-WIRED OR WIRELESS

Your decision to use wired or wireless alarm equipment will depend to some extent on your company's operations. Both have benefits, both have drawbacks, and both offer a variety of features to meet your requirements.

Do not overlook the recent advancements in radio frequency wireless technology, which now rivals hard-wired systems in features and performance. A detailed discussion of the new high-tech, supervised RF wireless equipment is included in Chapter 16.

FIRE ALARM CONTROLS

Selecting fire alarm control panels is similar in many respects to selecting intrusion alarm panels. The major difference is that local, state, and national ordinances and laws govern fire alarm installations. Requirements vary from state to state and from city to city. Because they are so different, attempting to address them all would make a book by itself. Included here are some general guidelines for your consideration.

Above all, the panel, as well as the rest of the equipment in fire alarm systems, should be approved by a recognized agency: Underwriters Laboratories (UL), Underwriters Laboratories of Canada (ULC), and/or Factory Mutual (FM). Additionally, if you install fire alarms in California, they must be approved by the California State Fire Marshal. Check your local and state laws to ascertain specific requirements.

General Considerations

Listed below are some general features that you should consider for a fire alarm control panel. The list should be used only as a guide.

- 24-volt operation
- Standby power supply with a regulated charging circuit and sealed, supervised, maintenance-free, rechargeable batteries
- Ground fault detection
- Lights (LEDs) showing AC power and trouble status
- Audible trouble signal
- Remote trouble signal capability
- Supervised alarm and protective circuits
- Supervised auxiliary power circuit for smoke detectors
- Auxiliary alarm output relay contacts
- Individual zone outputs (for multizone systems)

It is important to remember that your customer is depending on you, the quality of the equipment and your installation expertise to protect lives and property from fire. Plan and install fire alarm systems carefully.

Installation and Maintenance

All of the functions of an intrusion or fire alarm system culminate in the control panel. Therefore it must be installed and maintained properly for the system to perform properly. The new solid-state, microprocessor-controlled equipment requires special care and handling.

Reminding you to read the instructions before attempting to install a control panel (or any piece of equipment) may seem like unnecessary nagging. It's not. Minor design changes may not be readily apparent from the panel's physical appearance, but they could be significant in terms of system operation. Here are a few tips to help you avoid problems when installing control panels:

1. Test the unit. Hook up the control panel and test it on your bench and let it operate for at least a day before you install it.
2. Provide constant power. Make sure that the panel is connected to a twenty-four-hour power source, not one that will be inadvertently turned off causing the panel to switch to standby power.
3. Check all connections. Before applying power, make sure that all connections have been made correctly. Then check them again.

4. Verify primary power. Be certain that the transformer you are using is the proper size.
5. Provide independent power. Do not share the control panel's transformer with other equipment.
6. Use separate circuits. Keep power circuits away from protective circuits to prevent induced AC on your protective loops.

To prevent induced AC, radio frequency interference (RFI), and voltage transients, ground the panel. Use a 16-gauge or larger wire and connect the panel (according to the manufacturer's instructions) to an earth ground, like a cold water pipe. Do not use a gas pipe, electrical system, or telephone ground. Keep the ground wire as short and straight as possible. If you cannot verify a good earth ground to a cold water pipe, install a copper grounding rod yourself.

Detection Devices 14

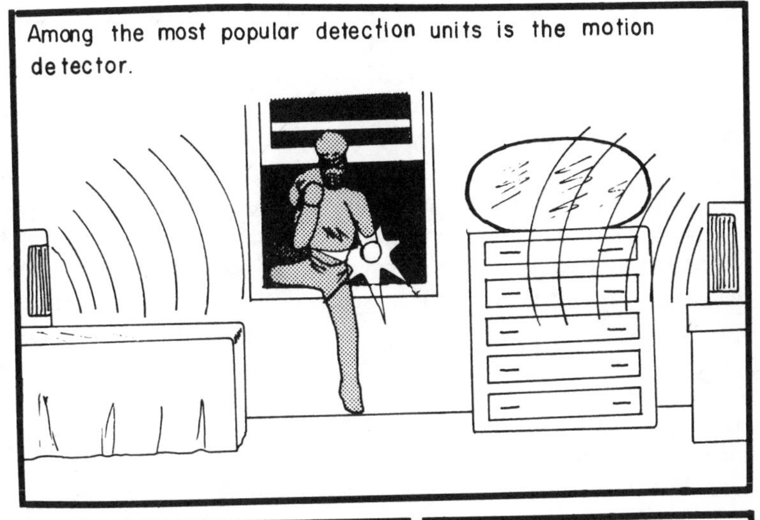

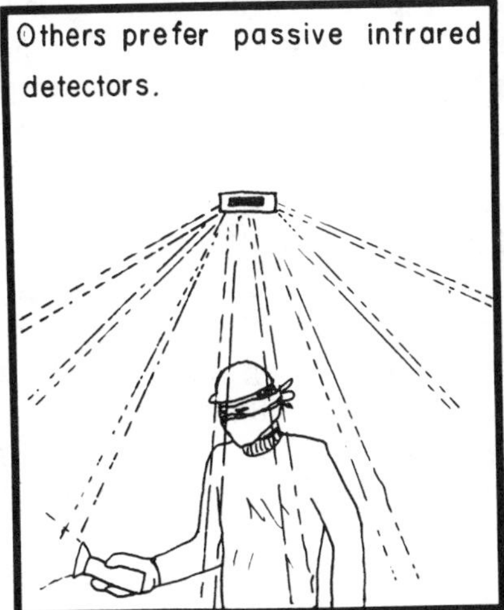

As noted in Chapter 12, a detector does just what its name implies—it detects something. A contact switch can detect a door or window opening. A passive infrared detector responds to an intruder's body heat. With the variety of devices that are available and a bit of ingenuity, you can detect almost anything.

In this chapter we take a look at most of the common intrusion and fire alarm system detection devices. Chapter 16 discusses specialized detectors and detection systems.

CONTACT SWITCHES

Contacts are such common items that they often do not get the attention that they deserve. They are available in myriad types, sizes, and electrical configurations. All you need to do is decide which ones meet your installation requirements.

Magnetic Contact Switches

Magnetic contact switches are available in two general types: mechanical and reed. Mechanical switches are composed of two metal strips enclosed in a plastic housing. When the magnet is moved away from the switch, the two strips, which were held together by magnetic force, separate, opening the normally closed circuit. The reverse occurs if the magnetic switch is designed for a normally open circuit.

Reed switches operate in the same manner. The metal contact strips are sealed inside a glass or plastic capsule that is housed in a plastic case. Although reed-type switches are not as susceptible to corrosion, they are much more delicate than mechanical switches.

Both recess- and surface-mounted switches have applications in alarm systems. Many dealers and installers prefer recess-mounted switches because they are concealed, making them less prone to tampering and making the installation more attractive. Figure 14.1A shows a one-quarter-inch diameter recessed switch and magnet. Others such as the one shown in Figure 14.1B are larger and are available for metal doors.

Sometimes it is not possible to used recessed switches, however, and surface-mounted models must be installed. These are available in a variety of sizes, shapes, and colors. A typical surface-mounted switch with attached leads is shown in Figure 14.1C. Usually, it is mounted on the door or window frame and the magnet is installed on the movable portion of the opening, either the door or the window.

Wide-gap switches can be used on loose-fitting doors. They can be installed up to two inches away from the magnet. Be sure to measure large gaps and select switches that will perform at those distances. Remember, too, that the gap distance listed on a switch's specification sheet is an "air gap." That is, it is the distance the switch will operate in free air. If you are installing the switch on a metal door, the gap will be reduced. Test the switch and the gap before you permanently install it.

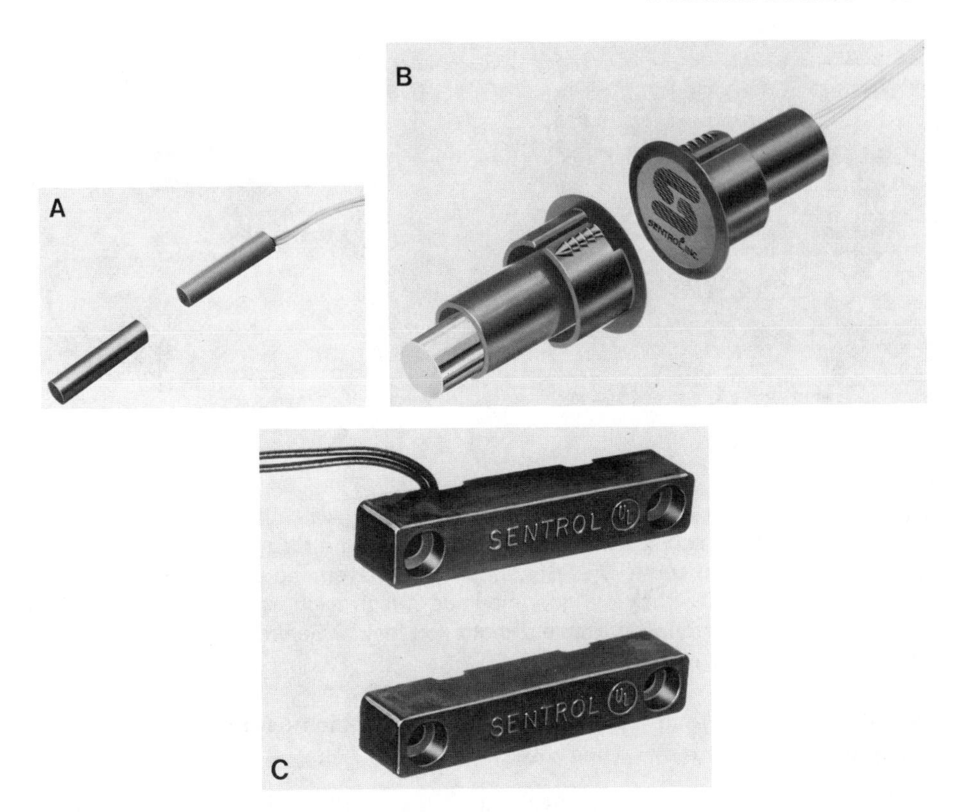

Figure 14.1 **(A)** Recessed switches and magnets help make a neat and attractive installation. Because they are concealed, they are more tamper resistant. (Photo courtesy of Ademco.) **(B)** Larger switches are sometimes easier to install. Since they are recess mounted, they are hidden from view and their size is not noticeable. The switch and magnet shown are available in models designed for steel or wooden doors. (Photo courtesy of Sentrol, Inc.) **(C)** Surface-mounted magnetic contact switches are available in several models. Some are prewired; others have terminals for attaching wires. (Photo courtesy of Sentrol, Inc.)

Special switches are designed for specific applications, for example, overhead doors. Overhead door magnetic contact switches can be floor mounted (Figure 14.2A) or mounted on the door's track with brackets (Figure 14.2B).

Another type of switch is specially designed for areas in which flammable or explosive conditions exist. Ordinary switches might cause a spark and ignite vapors, so explosion-proof switches are recommended. They are sealed, and the wiring is usually enclosed in conduit. In most instances, a switch of any type will not create a spark, but explosion-proof switches add an extra measure of safety.

Most magnetic contact switches are available for closed- and open-circuit protective loops. A few are single-pole–double-throw (SPDT) switches that offer

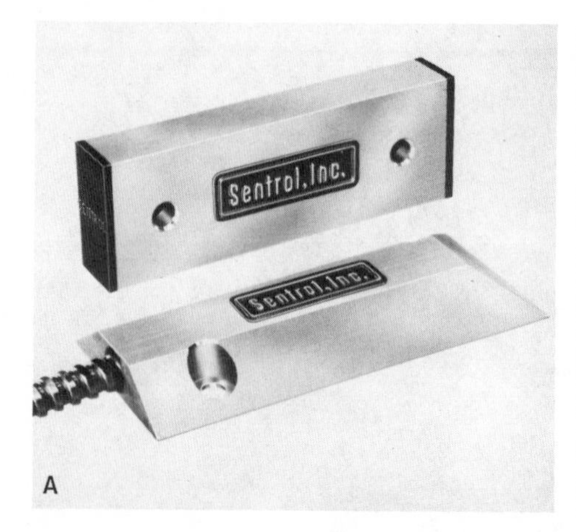

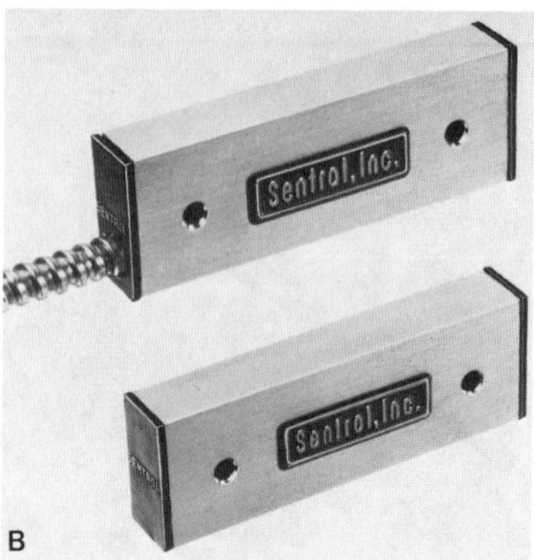

Figure 14.2 **(A)** This overhead door magnetic contact switch has a metal protective cable. The switch portion is mounted on the floor and the magnet is installed on the overhead door. (Photo courtesy of Sentrol, Inc.) **(B)** Overhead door switches can be mounted on the door's track if a floor-mounted switch might receive abuse. Special brackets make track mounting simple. (Photo courtesy of Sentrol, Inc.)

normally closed (NC), normally open (NO), and common terminals or leads, and can be used for various switching jobs.

Magnets. Magnets are not as rugged as you may think. Dropping one causes it to lose part of its magnetism, or pulling power. Repeated dropping or abuse may cause it to be only one-half as effective as it should be. A fully magnetized magnet is needed to perform efficiently in a security system.

Polarity is important, too. For the switch to work efficiently, the magnet must be positioned properly. Manufacturers design their magnetic contact switches differently, so it is a good idea to read and follow the installation instructions provided.

Installation Pointers. Here are a few suggestions for making magnetic contact switch installation easier and your alarm system more reliable.

1. When installing recessed switches, make the hole slightly larger than the switch.
2. Make sure that the switch and magnet are aligned properly.
3. Check gap distances before and after installation.
4. Check the protective circuit's resistance after installation (if necessary, check the resistance of each switch).
5. In damp environments, used sealed switches or apply a waterproof seal to the switch's terminals.
6. Do not overtighten screws.
7. Use spacers on metal doors and windows.

8. Add an extra adhesive (silicone rubber or epoxy) to self-adhesive switches and magnets.
9. Exercise caution when testing loop continuity. Some switches only have a rating of 0.5 to 1 amp. Check the switch's specifications and do not exceed the amperage rating listed.

Only infrequently should a magnetic contact switch cause system problems. High-quality switches may last up to twenty years if installed and maintained properly.

Application Notes. For several years, Sentrol, Inc. has published its booklet, "Application Notes." It describes various applications for switches and offers installation suggestions. The following is just a sampling of what you can protect with switches: art objects and paintings, truck tailgates, boat hatches, desk drawers, sliding doors and windows, overhead doors, attic hatches, fence gates, office machines, skylights, thin-framed windows, china and gun cabinets, casement windows, double-hung windows, Andersen windows, Pella windows, and revolving doors.

Ask your local distributor or call Sentrol, Inc. to find out how you can get your own copy of "Application Notes."

Other Contact-Type Switches

In addition to magnetic contact switches, other types of contact switches are useful to alarm installers. Undercarpet mats (switches) are used to protect areas where intruders are likely to walk; vibration contacts detect forced entry; plunger-type switches can be used on doors, windows, and other protection points as well as being installed on control boxes as tamper switches; mechanical "take-off" switches are used on sliding doors and windows, providing a take-off point for foil tape or other glass-breakage sensors to connect to the alarm system's protective circuit.

Most alarm systems use a variety of switches to meet customers' exact needs. Your use of switches is limited only by your creativity.

Plunger-type Switches. Plunger switches have many security system uses. Switches such as those shown in Figure 14.3 can be used to protect doors and windows or installed as tamper switches on control boxes. They also can be installed to protect gun cabinets, not only on cabinet doors, but recess-mounted in the base. The switch is depressed when a rifle butt is sitting on the plunger; moving the rifle activates the alarm.

Under-carpet Protection. Regardless of the nomenclature, under-carpet sensors, pressure mats, or switch mats do the same thing: when stepped on they trigger an alarm. Opinions about pressure mats vary; some installers like them and some don't.

The biggest complaint voiced by installers is that putting in the mats requires picking up and replacing carpeting. It is not the kind of task you should tackle unless you are good at it and prepared for it.

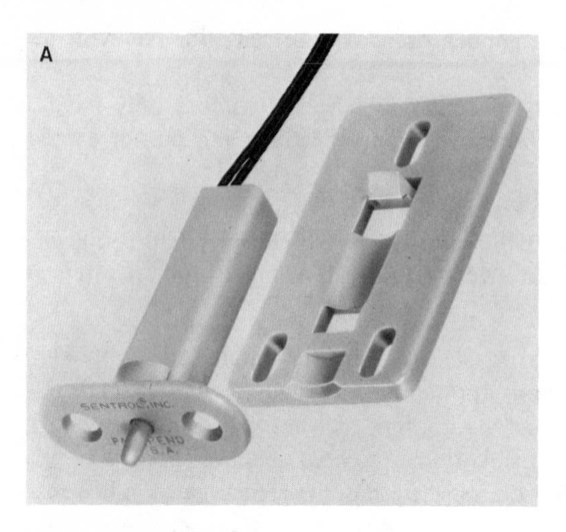

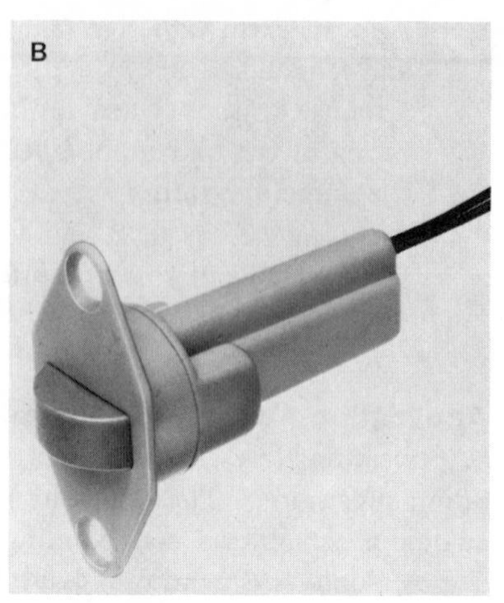

Figure 14.3 **(A)** Plunger-type switches have a variety of uses. They can be installed on cabinet doors, drawers, gun cabinets, and showcases as well as being used as tamper switches for control boxes. **(B)** The roller-plunger switch available in normally closed, normally open, and SPDT versions for maximum application flexibility. (Photo courtesy of Sentrol, Inc.)

Another problem with mats is pets. Some manufacturers have mats that are designed not to trigger an alarm when a pet walks on them. The pet's size and weight are critical factors, of course; and while a walking pet might not trigger an alarm, if that pet jumps off a sofa, the additional pressure might.

In some applications, however, pressure mats are ideal. They provide reliable spot protection where other devices simply would not work.

Before you embark on a program of installing them, check with other dealers who have used them or call your distributor. Removing and replacing carpet is time consuming and expensive, especially if you damage the carpet in the process.

Vibration Contacts. Vibration contacts are similar to magnetic contact switches because they have two metal contact strips housed in a plastic case. Instead of closing because of a magnetic pull, they close when the device is vibrated. One of the metal strips is weighted and its sensitivity is adjustable.

Vibration switches serve a useful purpose in some alarm systems, however, they must be applied and installed correctly to work properly. Installing them on windows is a less-than-ideal application. A better application is to install them on the window frame.

These devices are well suited to protecting interior and/or exterior walls to detect forced entry. Adjusting their sensitivity is critical. If they are too sensitive, they will trip the alarm whenever a large truck rumbles past or whenever a heavy thunderstorm hits. If they are not sensitive enough, they won't detect an intruder pounding on the wall.

Switches that react to vibration are usually fast-acting devices. Be sure that

Figure 14.4 Shock sensors usually cannot be connected directly to a wireless transmitter. Here, the sensor is connected to an interface, which lengthens the pulse of the fast-acting sensor so it will trigger the transmitter. (Photo courtesy of Litton Security Products.)

your control panel has either a fast-response circuit or a "pulse stretcher" on the protective circuit using the vibration contacts.

Shock Sensors. Instead of using metal contact strips, shock sensors detect impact and/or vibration in other ways. Common shock sensors use mercury, tuning forks, ring contact, or piezoelectric sensing methods. They tend to be more reliable and less susceptible to false alarms than vibration contacts. They are also more expensive.

With an interface or pulse stretcher, shock sensors can be used with wireless transmitters. Figure 14.4 shows a shock sensor connected to an interface and an RF transmitter. Note that it is mounted on the window frame for more reliable and efficient operation.

GLASS-BREAKAGE DETECTORS

Breaking glass can be detected by several types of products. Some are placed directly on the glass, some are not. The devices range from simple foil tape to sophisticated audio discriminators.

Foil Tape

Foil tape that is applied and maintained properly will give years of reliable protection. The key is to be sure that the foil adheres. The glass to which it is applied should be clean and dry. Once applied, it should be protected with a clear varnish or sealing material especially designed for that purpose.

Self-adhesive foil is easier to apply than the nonadhesive type, which requires an undercoat of varnish. Learning to install either kind takes time and a lot of patience. Novice installers should measure and mark the outside of the glass before installing the foil on the inside. A grease pencil and a template, which need be nothing more than a small block of wood, can be used. Then the foil is applied, following the line marked on the glass. In time, good installers use only their eyes and steady hands to apply foil tape. I know one who can apply foil in a circular pattern without marking the glass, a feat few of us can accomplish without using an installation aid.

A foiling tool is helpful to both novice and seasoned installers as it makes the job simpler and go more quickly. The Foilmaster has an adjustable arm that helps keep the pattern straight. It also has an optional template set for making round shapes. Large plate-glass windows can be foiled in minutes and the pressure pad on the tool helps create a tighter bond between the self-adhesive foil and the glass. Making good, neat square corners with foil tape requires practice, whether you are using a tool or are applying it by hand (Figure 14.5).

After doing it both ways, I much prefer the tool. With only a few exceptions, Sanger and Son, Inc.'s commercial customers could expect to have foil tape applied

Figure 14.5 The tool that makes these square and round corners in the foil's pattern is easy to use and allows you to apply the foil much more quickly than you could do it by hand. (Photo courtesy of Foilmaster, Inc.)

to their windows. Moreover, they could expect to have rounded corners in the foil's pattern. Because few other alarm companies in Oklahoma City used round corners, it almost became our trademark.

Dealers' opinions about glass protection vary. Beauty is in the eye of the beholder and when I step back from a well-foiled window, I think it is beautiful. Additionally, it serves as a visual deterrent for would-be intruders.

The chief complaint most dealers voice about foil is that it is expensive to install and maintain. The foil itself is not expensive, but the labor to install it can be. Also, if it is not installed and maintained properly, you can almost count on numerous service calls. My experience with foil is that it is not expensive to install because the foiling tool significantly reduces application time, and because we took special care to protect the foil with a heavy coat of varnish, few service problems resulted.

Actually, foil is one of the simplest and best protective devices there is. It works—if the glass breaks, so does the foil tape. It is effective. It is either in one piece or it isn't, and it rarely causes false alarms. Occasionally, a hairline crack may cause a swinger (an intermittent, difficult-to-trace alarm signal). If the alarm system is zoned, isolating the source of the false alarm should not be difficult.

Taking it Off. Differences of opinion exist among alarm installers who regularly use foil tape with respect to the best method for connecting the foil to the protective circuit. A variety of take-off devices are available; some of which require soldering, others do not. I have used simple, self-adhesive take-off blocks with screw terminal clamps for years without problems. Occasionally, a take-off block would come loose or get knocked off the glass. Usually, however, these problems were reported by the customer or noticed during a regular maintenance inspection before an emergency service call was required or a false alarm was caused.

On movable windows, special take-off devices are required. The one shown in Figure 14.6 can be used on windows or sliding glass doors.

Repairing Foil Tape. Foil tape repairs can be simple:

1. Clean the glass around the broken tape.
2. Scrape away the varnish with a razor blade and/or steel wool (down to where the foil has a shiny appearance).
3. Apply a strip of foil tape over the break.
4. Press the new tape firmly over the broken tape.
5. Using a knife point, small screwdriver tip, or other sharp instrument, press the new tape onto the old tape (this helps ensure good contact).
6. Apply a heavy coat of varnish over the repaired section.

At some point, you should consider refoiling the glass instead of repairing damaged tape. Numerous splices and patches make the job unsightly. As a general rule, if it has been repaired three times, refoil the glass.

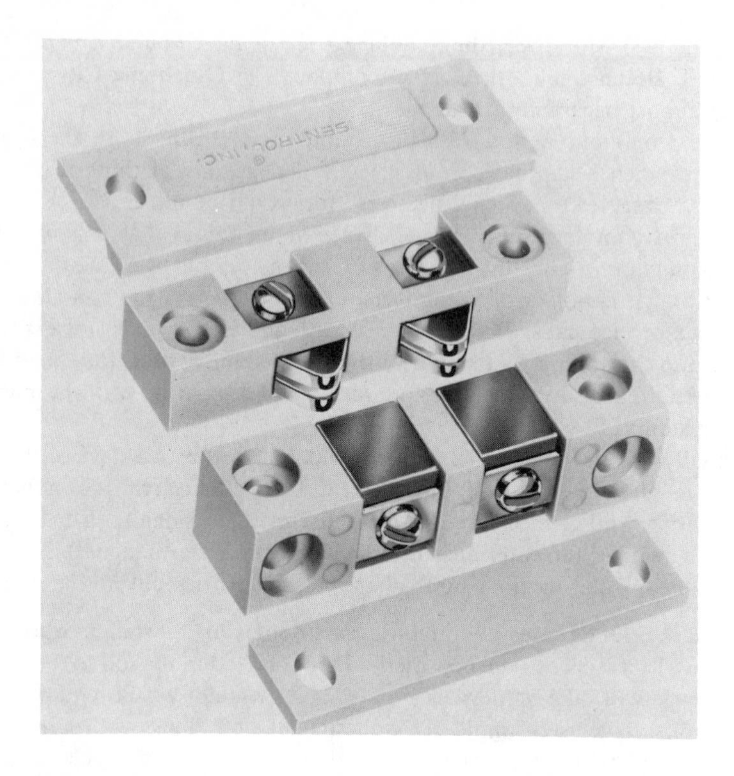

Figure 14.6 Sliding doors and windows pose special problems for connecting foil to protective circuits. Take-off devices, such as the one shown, help make good, simple connections. (Photo courtesy of Sentrol, Inc.)

Other Protection Methods

Other devices to detect breaking glass include audio discriminators, which are discussed in more detail later in this chapter, and shock-type sensors applied directly to the glass. These respond to the frequency of breaking glass, not to vibration. Both types of sensors work very well when applied properly and are ideal for those installations for which foil tape is not acceptable.

ALARM SCREENS

Alarm screens are becoming increasingly popular with alarm dealers and customers. Protection and ventilation are provided, the system can be armed with the windows open, and the homeowner can still be assured of complete protection. Basically, they are ordinary window screens with small, insulated wires laced into the fabric. Manufacturing methods differ, so minor differences in construction exist. Check various manufacturers' products to find the one best suited to your installation needs.

Whether you want to use alarm screens as the only detection device for windows will depend on your philosophy of security protection. You may consider using magnetic contact or other switches on the window for back-up protection. The window devices could be included on a separate zone so your customer could shunt that zone, open his windows, and then arm the system. When the house is unoccupied, both the alarm screens and the window contacts would provide security.

Alarm screens are available for most types of windows. Sliding and swinging screens, however, like those found on front, back, and patio doors, tend to cause problems because of their frequent use.

Another consideration: beware of cats. Much to the chagrin of their owners, cats occasionally like to climb screens. If the cat's claws snag one of the wires, you will have an activated alarm and a service call to make to repair or replace the screen.

Installation and equipment costs for alarm screens may run two to three times more than those for more conventional window protection. A large number of windows will quickly increase the price of the system.

SPACE PROTECTION

Space protection is known by a variety of terms; volumetric security, area protection, and motion detection are just a few. Whatever you call it, it is an efficient and effective way to protect large spaces.

Five types of detection devices are included under the heading of space protection: ultrasonic and microwave motion detectors, passive infrared (PIR) detectors, photoelectric beams, and audio discriminators. It is unlikely that I will get any arguments against including the first three. My rationale for including the last two is that photoelectric beams, although often used for perimeter protection, actually protect long, slender spaces; audio discriminators "listen" for certain sounds that could emanate from anywhere in the protected area; and both devices protect volumes of space instead of just a specific point.

If you count combination detectors, or dual detection devices, that contain two different detectors in a single housing, you have a sixth category. Only a few devices of this type are currently available, however.

Previously, space-protection devices were used to protect only high-risk customers, places that might be targets for professional burglars. Now, many dealers routinely install them, as they can be used either to protect larger areas instead of specific points or to detect stay-behind criminals. Equipment costs have decreased as the devices have become popular and are used more frequently. Chances are that when newer units are introduced, they will be less expensive and more reliable because of technologic advancements.

When planning an installation that will use one or more space-protection devices, think like a thief. Try to determine what areas he would most likely enter; those are ideal areas in which to place space-protection devices.

Of course, further consideration must be given to the environment in which those units will be used. Each type of device has advantages and disadvantages. Understanding their application will help you provide better protection for your customers and reduce installation and service time, as well as inadvertent alarms.

Motion Detectors

Ultrasonic and microwave detectors respond to motion. Some call passive infrared detectors motion detectors, too, but they actually sense thermal changes, not movement.

Ultrasonic Motion Detectors. Ultrasonic detectors use transmitters and receivers that broadcast and receive high-frequency sound inaudible to humans. The units are available as transceivers, with the transmitter and receiver housed in the same cabinet, or with transmitting and receiving separate. Both work well if installed properly; the transceiver unit is simpler to install, however.

The sound waves emitted from an ultrasonic transmitter will not penetrate most materials, making its protection pattern easy to define. Generally, the sound waves are blocked by any solid material. Soft materials, like fabric, may allow some of the sound waves to pass through them; some of the waves will be blocked and the unit's range reduced, however.

The environment in which an ultrasonic detector is installed affects its range and performance. Soft environments with heavy carpets, draperies, and upholstered furniture will absorb sound waves and reduce the unit's range. Hard environments with tile floors, brick walls, and wood or metal desks tend to reflect the sound waves and may actually increase the detector's range. In some hard environments, I have observed some ultrasonics' sound waves bounce around corners. It can happen, but don't count on the bounce and extended protection area.

Humidity affects ultrasonics. As humidity increases, sensitivity and range decrease, and vice versa. Range also may increase a bit after your customer leaves the building and the internal air turbulence stabilizes.

Air turbulence can affect an ultrasonic's operation and excessive amounts may cause the unit to trigger an alarm. Most newer ultrasonics have special circuits to minimize alarms from air turbulence. They should be installed at least ten feet from heating and air-conditioning vents or any other sources of air turbulence, however.

Certain high-pitched noises, like bells and squeaking machinery, may cause problems. Before installing an ultrasonic, check for possible noise sources and take steps to eliminate or minimize them.

Ultrasonic detectors usually have a maximum range of about thirty square feet depending on the environment in which they are installed. Because they flood the area with sound waves, a small office, for example, would have wall-to-wall and floor-to-ceiling protection.

If you will be installing ultrasonics in several areas, you should consider master-slave systems instead of individual, stand-alone transceivers. Master-slave systems usually have a lower overall cost and provide as much protection as individual units. Slave detectors are powered and controlled by a master unit, which may or may not be a detector itself.

Following the instructions provided by the manufacturer is very important if you want to minimize installation and service time and false alarms. Resist the temptation to make the unit oversensitive or extend its range beyond what the manufacturer recommends. Other installation tips are as follows:

- Be sure that the ultrasonic's signals are not blocked by furniture or merchandise.
- Do not install the unit more than twelve feet above the floor.
- Adjust the system's range to cover the desired area. Do not try to overextend its range.
- Avoid sources of excessive air turbulence.
- Identify and correct sources of high-pitched noises like telephones, bells, squeaking machinery, or hissing air compressors. If the noises cannot be eliminated, keep the ultrasonic device as far away from these noise sources as possible.
- Be sure that pets will not be allowed to roam through the protected area.
- Check for any other items that may move in the protected area, for example, curtains and hanging signs.
- Mount the device on a vibration-free surface.
- If more than one ultrasonic unit will be used in the same area, be sure they do not "talk" to each other (check with the equipment manufacturer about eliminating this problem).
- Use test meters to help ensure that the system is installed properly and is covering the desired area.

In some instances, an ultrasonic detector may interfere with other devices, for example, if it is on or near the same frequency as a remote-controlled television. I know of one installation where the ultrasonic motion detector, once in place and operating, changed the channels on the customer's remote-controlled television—whether the set was on or off. It is a good idea to check for ultrasonically controlled devices before installing the detector.

Ultrasonic detectors are available in many shapes, sizes, and colors. The unit shown in Figure 14.7 can be mounted on the ceiling or wall. The ones pictured in Figure 14.8 are available in white and brown wood grain and can be installed on a wall or corner.

Microwave Motion Detectors. In physical appearance and general operating characteristics, microwave detection units and ultrasonics are similar. Some significant differences exist, however.

Microwave detectors operate at a much higher frequency than ultrasonic

Figure 14.7 Some ultrasonic detectors are compact, attractive devices. They are cost effective, reliable, and easy to install. (Photo courtesy of Raytek, Inc.)

detectors, and their energy penetrates most nonmetallic objects. In a sense, they can be compared to radar.

Although the energy will penetrate nonmetallic objects, the amount of penetration is determined by the composition of the penetrated material. The denser the material, the less penetration. Because of this, proper installation is critical. For example, if you point the unit at a window it will detect movement outside and possibly cause a false alarm.

Figure 14.8 Many units are available in attractive housings for residential use. The wood grain appearance of the detector shown here allows it to blend well in paneled rooms. (Photo courtesy of Arrowhead Enterprises, Inc.)

Protection patterns are more varied than those of ultrasonic detectors. A microwave's pattern may extend 300 feet or more on some models, and it can be wide or narrow.

Air turbulence does not affect microwaves as much as it does ultrasonics; however, reflected signals can be a problem. Large metal surfaces act like mirrors, bouncing the energy back into the microwave's antenna or into an area that is not part of the protected area.

Fluorescent lights may cause interference problems, too, and microwave units should be placed at least ten feet away from the nearest one. Ideally, these lights should be turned off when the microwave is part of the protection system.

Other installation tips include the following:

- Avoid pointing the unit directly at large metal objects, like overhead doors, to minimize reflection problems.
- Check adjacent areas to be sure that the microwave's signals do not penetrate windows and walls.
- If using two or more microwaves in the same area, be sure to use different frequencies to avoid interference problems.
- If you are counting on multiroom coverage based on a microwave's penetration capabilities, be sure that the wall you are attempting to penetrate does not contain a lot of metal.
- Mount the unit on a vibration-free surface.

Causes of problems are not always obvious. One alarm system that included a microwave detector continued to have false alarms. It was discovered that the system would trip whenever anyone walked on the porch—but the microwave was not aimed at the porch. When the problem was solved, it turned out that the microwave was aimed at a wall that was covered with metallic wallpaper. The microwave's energy was being reflected off the paper and through a window onto the porch.

A microwave's energy is designed to detect a man-sized object at its rated distance. For example, on a fifty-foot model, it should see a man walking at any distance up to fifty feet away. Its energy does not stop at its rated distance, though. It can "see" larger objects at greater distances. If it is designed to see a man at 50 feet, it also could see a car (because of the car's large metal mass) at 100 feet or more.

Overall, microwave detection devices tend to be stable, reliable, and effective. Apply and install them properly and you will have few problems.

Passive Infrared Detectors

Passive infrared (PIR) detection devices are not new, but they have only recently increased in popularity with alarm installers.

These devices react to temperature changes, which they look for by sensing infrared energy emitted and reflected by objects within their fields of view. Because they are "looking" instead of transmitting energy, they are considered passive devices.

A simple explanation of how a PIR works is that it monitors the radiant energy within the protected areas and compares those readings to any changes in the infrared environment. If the changes in temperature are rapid or exceed certain preset limits, the PIR triggers an alarm.

All objects transmit infrared radiation to some degree. For that reason, some PIR detectors require a heat source to be in motion before tripping the alarm. Others have segmented protection patterns and require that two or more protection elements sense thermal changes before going into an alarm condition.

Many PIR models can be aimed with accuracy so that exact areas can be protected. By keeping a PIR's protection pattern several feet above the floor, pets can roam undetected. (If pets jump or climb, as cats are wont to do, a PIR will detect them, of course.)

The PIRs have protection patterns that extend from a few feet to more than 100 feet. These patterns can be wide or narrow and can be solid or segmented.

Manufacturers have designed PIRs to be aesthetically pleasing. Some have attractive housings (Figure 14.9A) that can be surface mounted without detracting from a room's decor. One model is designed to look like an air vent (Figure 14.9B) and another appears to be an electrical outlet (Figure 14.9C).

Installing PIRs is no more difficult than installing other types of space protection devices. Here are some installation suggestions:

- Point PIRs at objects (e.g., walls or floors) within their rated ranges. Do not have them looking off into infinity.
- Make sure their reference point (the wall or floor noted above) has a stable temperature and does not warm or cool rapidly.
- Mount PIRs away from air-conditioning and heating vents.
- PIRs should not have any objects that heat or cool rapidly (bare light bulbs, fireplaces, windows) in their fields of view.
- Avoid mounting PIRs where direct or reflected sunlight or car headlights can strike them.
- Mount the units on a firm, vibration-free surface.

A major benefit of PIRs is their immunity to disturbances that cause other space-protection devices to trigger false alarms. They must be applied and installed properly for optimum operation, however.

Photoelectric Beams

Although photoelectric beams are available with visible and invisible light sources, the latter are more useful in alarm systems. Most alarm system photoelectric beams,

Figure 14.9 **(A)** Some PIRs have decorator-styled cases, making them attractive as well as functional. (Photo courtesy of Aritech Corp.) **(B)** To make them even less noticeable, some manufacturers disguise their PIR detectors. This unit, once installed, looks like an air vent. (Photo courtesy of Raytek, Inc.) **(C)** Disguised as an electrical outlet, this passive infrared detector provides protection without detracting from a room's decor. (Photo courtesy of Detection Systems, Inc.)

usually referred to as just "beams," use a pulse-modulated infrared light beam. Breaking the invisible beam triggers an alarm.

There are two varieties of beams. One uses separate transmitters and receivers. The other uses a bounce-back technique, and the transmitter and receiver are housed together. For long-range applications, individual transmitters and receivers are required, leaving the bounce-back devices for short-range applications.

Long areas, like hallways and warehouse corridors, can be economically protected by photoelectric beams. Some models are capable of protecting areas up to 1,000 feet long.

Alignment is critical. You must be sure to have the transmitter and receiver aligned exactly. The greater the distance between the transmitter and receiver, the more difficult alignment becomes. Beams can be surface mounted, such as those shown in Figure 14.10, or they can be flush mounted, with some units disguised as electrical outlets, similar to the one shown in Figure 14.9C.

Whether surface- or flush-mounted, photoelectric beams should be installed on solid, vibration-free walls. If the units are surface mounted, they should be protected so that they will not be accidentally bumped and knocked out of alignment or broken.

Audio Discriminators

Sound detectors, audio discriminators, sonic detectors—whatever you want to call them—have been around for a long time. They have been used in alarm systems for a long time, too. Although dealers' and installers' opinions about audio discriminators vary greatly, the new audio technology is proving that they are economical and effective devices.

Here is a simplified explanation of how an audio sensor works:

1. A microphone picks up sound.
2. The sound's intensity (decibels) and frequency (vibrations per second) are converted into an electrical signal.
3. The sensor's circuitry measures the electronic signal.
4. If the signal is within the sensor's acceptable limits, no alarm is triggered; if it is, an alarm is initiated.

Sound waves from breaking glass, metal-to-metal contact, and breaking or splintering wood are distinctive. True audio discriminators respond only to those types of sounds.

One benefit that results from using audio sensors is that a single detector can protect several windows, so they are cost effective. Instead of the high labor costs associated with protecting individual windows with foil, glass-breakage detectors, or magnetic contact switches, a single audio discriminator can be installed quickly and inexpensively.

Figure 14.10 Pulsed infrared photoelectric beams provide invisible protection for long, narrow areas like hallways and aisles. Beams are available in surface- and flush-mounted models, some having a range of 1,000 feet. (Photo courtesy of Detection Systems, Inc.)

Audio sensors can be surface or flush mounted on walls or ceilings. Surface-mounted models are available in several colors to blend in with most decors.

Testing audio sensors occasionally has presented installers with problems. Some manufacturers provide testing devices, but one of the simplest and most effective is a homemade tester: a sixteen-ounce nonreturnable soft-drink bottle containing six glass marbles. Shaking the bottle with the marbles simulates the sound of breaking glass.

When adjusting the sensitivity of each audio sensor, use the tester to help you calibrate the device correctly. Stand by each point you want protected and rattle the marbles in the bottle. If the sensor trips, you know its range. Also, if heavy curtains cover windows to be protected, shake the bottle behind the curtains to determine how much of the sound will be absorbed.

Some testing devices are made from metal or plastic strips. Snapping the strip against the window is supposed to simulate breaking glass. Of course, if you snap it too hard, you will have a real opportunity to test the detector because you will break the window.

Audio sensors, like all other detection devices, have limitations. They should not be used in areas where the sound of breaking glass might be muffled by other loud or high-frequency noises. Also, breaking glass sounds like breaking glass, regardless of what breaks. The audio sensor cannot tell the difference between a drinking glass dropped in a sink and the front window breaking.

Loose window panes also can cause false alarms. As the panes vibrate, they are actually grinding against their frames, and this grinding noise, though slight, is caused by small pieces of breaking glass.

One of the most important considerations in using audio sensors is to be sure that all windows lock securely. You want to force an intruder to break a window to gain entry, and breaking the window will trigger the alarm. If a window can be forced open without breaking its glass, an alarm may not be triggered.

Combination Detectors

Recently, some manufacturers have been combining detection technologies to effect reliable security devices. For example, ultrasonic and PIR detectors have been combined, as have microwave and PIR devices.

Dual-technology detectors will trigger an alarm only if both are activated. For example, a combination microwave-PIR must detect motion (by the microwave) and thermal change (by the PIR) at the same time. Hence it will detect an intruder, but not a book falling off a shelf.

Combination detectors tend to be more expensive than traditional single detectors. Their reliability and immunity to inadvertent alarms make them worth the higher cost, however.

Outdoor Protection

Many of the types of devices used indoors can be used outdoors, too. Microwave, photoelectric beams, and shock sensors are common outdoor protection devices, for example. Chapter 16 includes a detailed discussion of outdoor security devices.

Hold-up Alarm Devices

Basically, hold-up alarms are single-purpose devices: they are designed to respond to the manual activation of some type of switch. They are usually activated by a switch's contacts closing. Because the types of switches used in hold-up alarms are specialized, they are discussed fully in Chapter 16.

FIRE DETECTORS

Specialized, commercial-type fire alarm systems and detectors are beyond the scope of this book, however, the three most common fire alarm devices are discussed briefly because they are frequently used in residential and/or commercial systems not requiring stand-alone fire alarms.

Smoke Detectors

Small, battery-operated smoke detectors are common sights in most homes. They are being connected to professionally installed alarms more and more frequently as people realize the need to alert emergency and/or fire-fighting personnel to save lives and property. While a small battery-powered smoke detector will alert a homeowner and his family, it will not help dispatch assistance. Also, if no one is home to hear the small detector, a large fire could result before neighbors see it and call the fire department. To install smoke detectors that will provide the most protection for your customers, you should know their operating characteristics.

Ionization Detectors. Ionization detectors should be used in areas where fires might develop gradually, emitting invisible products of combustion first, then visible smoke, and finally flames. This type of detector responds quickly to invisible particles of combustion during a fire's earliest stage.

Dual-chamber ionization detectors are recommended. An outer chamber allows air, smoke, or invisible combustion gasses to enter freely; an inner chamber is nearly closed. When invisible products of combustion or smoke enter the outer chamber, an electrical imbalance is created between the chambers, triggering an alarm.

Ionization detectors are most responsive to class A fires involving wood, paper, and rubbish and class C fires such as those involving electrical equipment. In some instances, ionization detectors can detect equipment overheating before a fire starts and/or the equipment fails.

Photoelectric Detectors. Photoelectric detectors also sense fires in their beginning stages. They do not detect fires quite as rapidly as ionization detectors because they operate on a light-scattering principle.

A light source and light-sensitive receiver, like a photodiode, are located in the sensing chamber. When smoke enters the chamber, light is reflected from the smoke particles onto the photodiode, causing an alarm.

Photoelectric smoke detectors should be used in areas where it would be impractical to use ionization detectors, for example, where fumes or vapors are normally present or when the material expected to burn will produce visible smoke.

Thermal Detectors

Thermal detectors, also known as heat detectors, thermostats, or just "stats," are simple and reliable. They are available in fixed-temperature, combined fixed-temperature and rate-of-rise, and rate-compensated models.

Heat detectors can be used almost anywhere and they may be your only choice if an environment is not suited to smoke detectors. Because they sense only heat, they provide detection with a minimum of inadvertent alarms. A drawback is the time it takes for thermostats to trigger an alarm when a fire occurs.

Flame Detectors

Instead of detecting smoke or heat, flame detectors sense the presence of an actual flame through the flame's emission of ultraviolet or infrared energy. These are highly specialized devices with limited applications.

Ultraviolet flame detectors are highly sensitive to the ultraviolet portion of the energy radiated by all types of flames, including those produced by flammable liquids and gases. Some models can be used outdoors and are not affected by direct sunlight or intense lighting. They are ideally suited to protect industrial plants and oil rig platforms. The ultraviolet unit is sensitive to welding torches, and care must be exercised when installing these types of detectors in industrial areas.

Infrared flame detectors respond to a narrow band of infrared radiation that is present in hydrocarbon fires. Usually, infrared flame detectors are limited to indoor use.

INFORMATION SOURCES

Dozens of new detection devices are introduced by manufacturers each year. Caution is called for in taking a step from the familiar to the new and untried. New products do not always meet dealers' expectations, but new technologies promise many benefits. Most new alarm devices are more reliable and offer new or improved features.

Manufacturers, distributors, associations, and other dealers are good sources of information about new detection devices. Each source should be explored when you are considering new products. Some manufacturers and a few distributors offer evaluation programs for new products, allowing you to test and evaluate the product for a month or so before purchasing it. It could be worth while to explore these programs.

Annunciation Devices 15

The entire purpose of an electronic protection system is to sound an alarm. Webster's *New World Dictionary of the American Language,* Second College Edition, defines alarm as: "1. (Archaic) a sudden call to arms. 2. a signal, sound, cry, etc. that is a warning of danger." An annunciation device provides the warning and is the alarm portion of a security system. In other words, it announces an emergency or problem.

Annunciators can sound an alarm locally or at a remote location, they can be audible or visible. When most people think of sounding an alarm, they think of a bell. Other audible sounding devices include buzzers, sirens, and horns. A flashing light may be an alarm, for example, a floodlight, strobe light, or a tiny light-emitting diode (LED).

An alarm also may be silent. That is, there may be no indication on the protected premises that an alarm has been triggered. Rather, it is annunciated at a remote location—down the street, across town, or across the country. Often, alarm dealers provide their customers with both types of annunciation: a local sounding device with remote monitoring.

LOCAL ALARMS

Local alarms serve two purposes: (1) they notify intruders that they have been detected and (2) they notify neighbors and passersby that something is happening. Personally, I think the first purpose is more important, especially, because you cannot always count on the second. Few intruders are going to stay around once a bell or siren sounds. Besides, they won't know whether or not a neighbor has called the police, and it is unlikely they would wait to find out for sure. Moreover, for all they know, the alarm is ringing at the police station, too.

Some alarm dealers prefer local alarms; other prefer silent, remote alarms. I suspect that most prefer alarm systems to make noise at the protected facility and notify their central stations or monitoring facilities, as I do. If you have decided that making noise is a good idea, you have several alternatives.

Bells

Bells probably are the most common alarm sounding devices—they have been used for decades. Unfortunately, because they have been around for so long, people may ignore them. They are loud and they are easy to install, and some dealers wouldn't think of installing anything else.

Most alarm system bells operate on 6 VDC or 12 VDC. If your alarm control panel has a voltage output, chances are you can connect a bell to it. Check the panel's instructions first, however. Because bells may generate voltage transients, you may need to add a general-purpose diode, such as the one shown in Figure 15.1.

If a bell is installed on the exterior of a building, it should be protected

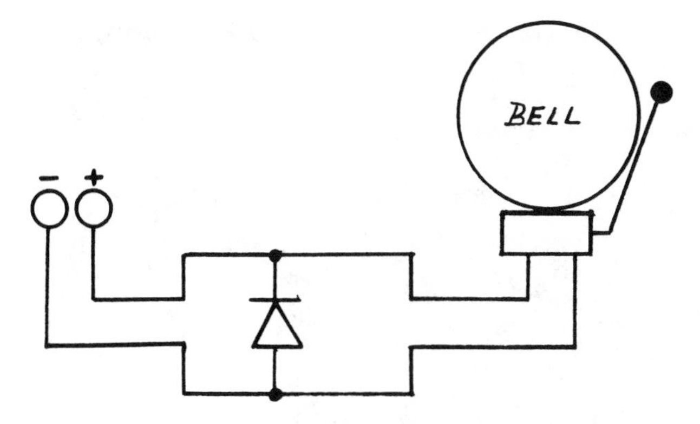

Figure 15.1 A general-purpose diode, like an IN4001 or equivalent, should be used across a bell circuit. Bell coils may generate voltage transients or spikes that could damage sensitive alarm equipment.

both from Mother Nature and from people intent on tampering with it. A metal enclosure, or bell box, accomplishes the former by keeping rain, snow, birds, and rodents away from the bell. One or more tamper switches attached to the bell box help prevent tampering.

To create a distinctive sound with a ringing bell, some control panels have a pulsed voltage output. This causes the bell to ring for a short period, then stop, then ring again, and so on. The pulsing bell probably attracts more attention than a steady ringing sound.

Sirens

If you want louder sounds and a greater variety of them, consider installing a siren. Of course, if it will be installed on the outside of a building, it should be protected.

Mechanical and electronic sirens are used in alarm systems, with the latter being more common. Mechanical sirens force air through casing vents with motor-driven impellers. Electronic sirens create similar sounds with oscillators, power amplifiers, and loudspeakers.

The oscillator-power amplifier unit usually is an electronic circuit called a siren driver, which can be a separate device (Figure 15.2A) or built into a loudspeaker (Figure 15.2B). Siren sounds include steady tones, wails, and yelps. Often, siren drivers such as the one shown in Figure 15.2C provide more than one type of sound and have adjustable ranges and pitches.

Connecting a single speaker to a siren driver is a simple task. One end of a pair of wires attaches to the siren driver, usually to terminals labeled "speaker," and the other end to the loudspeaker. If you are connecting more than one speaker

A
2 AMP.
J1
U2
500
SIREN
DRIVER
+
−
+
STDY
SPEAKER
4 OHM MIN.

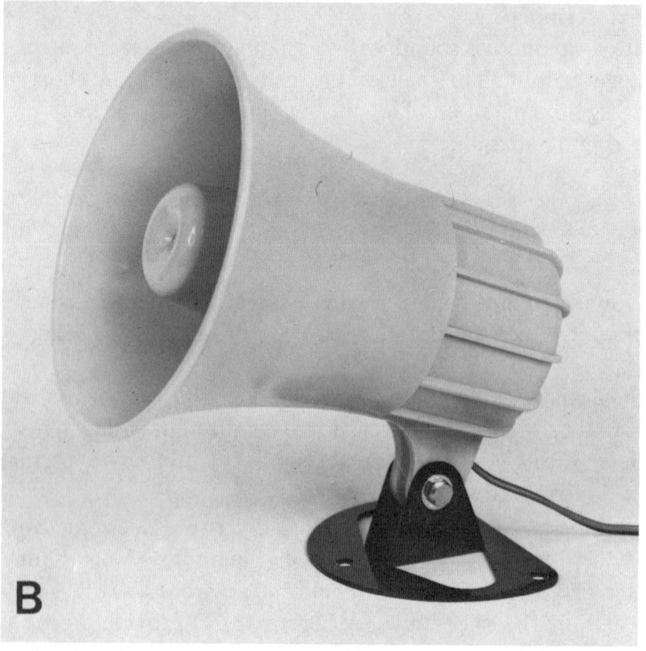
B

Figure 15.2 **(A)** Voltage output from a control panel triggers a siren driver. The oscillator-power amplifier circuit generates the sound, which is emitted through a loudspeaker connected to the driver. (Photo courtesy of AES Corp.) **(B)** Some siren drivers are built into speakers. The siren/speaker's two wires are connected directly to the control panel's voltage-output terminals. Larger combinations usually have four-conductor cables, with two of the conductors providing a tamper circuit. (Photo courtesy of American Security Equipment Co./AMSECO.) **(C)** Siren drivers like Adcor's SD-7 provide steady and yelp sounds in addition to adjustable rates and frequencies. Speaker-connection terminals on most siren drivers are clearly marked, but it is a good idea to read the instructions before making any connections. (Photo courtesy of Adcor Electronics.)

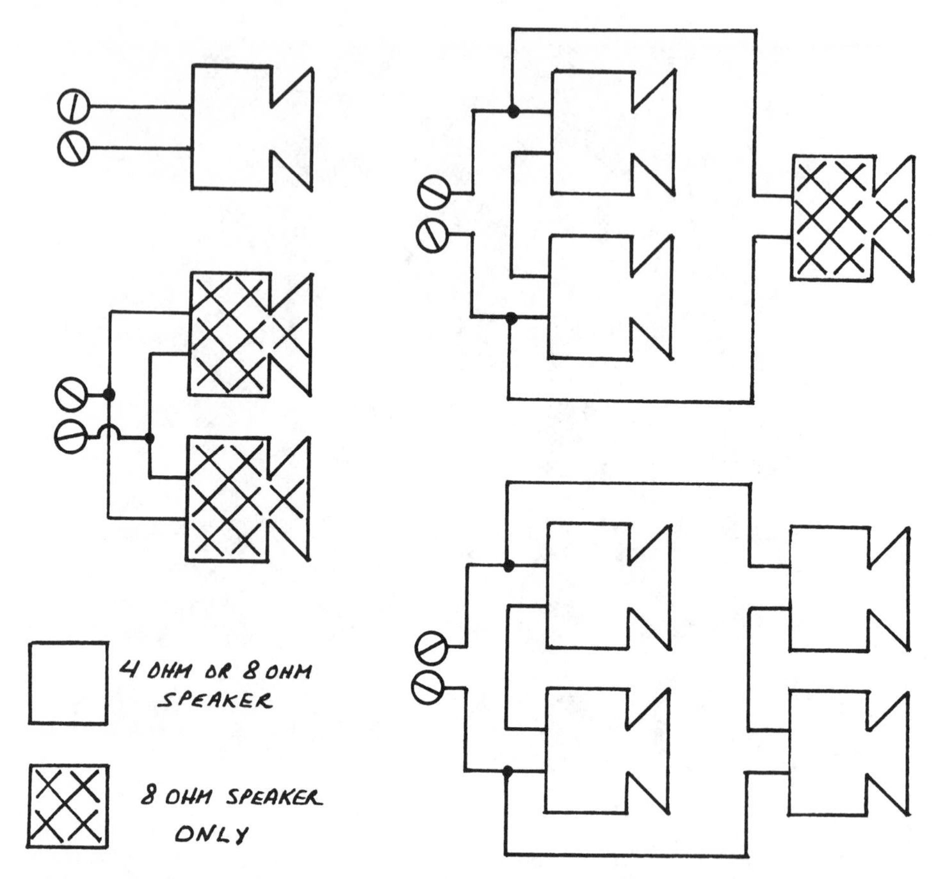

Figure 15.3 Care must be exercised when connecting more than one speaker to a siren driver. Read the siren driver's instructions carefully to determine whether 4-ohm and/or 8-ohm speakers may be used. Incorrect connections can damage siren drivers.

to a siren driver, you must exercise care. Figure 15.3 shows how to connect several speakers to a siren driver. Incorrect connections could damage the driver.

Speaker size affects sound output. The larger the speaker, the louder the siren. Speaker sizes are measured in watts and ohms (resistance). Thirty-watt, 8-ohm speakers are recommended for most applications to maximize sound output from a siren driver. When mounting speakers on the outside of a building, be sure to protect them from birds, the weather, and unauthorized tampering.

The wire connecting the loudspeaker to the siren driver may affect loudness, too. Use 14- to 18-gauge wire and keep wire runs as short as possible.

Sound Levels. Siren outputs, sound levels, are measured in decibels (dB). Most alarm sirens are rated from 90 to 120 dB and are usually specified at a distance

Table 15.1 Decibel Ratings at Different Distances

*10**	*20*	*40*	*80*	*160*	*320*	*640*
120	114	108	102	96	90	84
110	104	98	92	86	80	74
100	94	88	82	76	70	64
90	84	78	72	66	60	54
80	74	68	62	56	50	44

* Distances in feet.

As distance increases, sound level (dB) decreases. For comparison, a whisper at three feet is about 45 dB.

of ten feet. For example, a siren's specification sheet might read: "Output: 110 dB at 10 ft."

Decibels are calculated logarithmically. As a general rule, 6 dB are lost every time the distance doubles. For example, a siren rated at 100 dB at ten feet would have ratings of 94 dB at twenty feet and 88 dB at forty feet. Table 15.1 provides decibel ratings at various distances. Although a siren may sound loud at 10 feet, it may be difficult to hear at 320 or 640 feet.

Obstructions and breezes affect sound, too. You may have to experiment with several sirens and adjust their rates and pitches until you find one that gives the greatest sound output that will be carried over specific distances.

Interior Sirens

This is another point on which alarm dealers disagree. Some like to install sirens indoors and outdoors, others only outdoors, and a few only indoors. I prefer installing at least one both inside and outside. To my way of thinking, the more noise, the better.

Installing a 30-watt loudspeaker inside a factory probably will not affect the room's aesthetics. Installing one in someone's living room is a different story. You probably will have to opt for a smaller siren (speaker) to use inside homes. Several unobtrusive types are available for residential installations.

ZONE IDENTIFICATION

Another type of local annunciator is not for the purpose of frightening an intruder or signaling an emergency. Rather, it is to help you and your customer determine the source of an alarm.

Zone locators, or annunciators, help pinpoint trouble spots and alarm locations by identifying the exact zone in which an alarm was triggered. They offer numerous features that allow you to meet specific zoning requirements. Some can detect and annunciate swingers (brief, randomly occurring detector activations) without triggering a full alarm. Locating and correcting swinger problems helps reduce the number of false alarms.

Other features to consider include adjustable loop response, protective circuit supervision, automatic arming, individual zone outputs, and selective and/or automatic zone shunting.

Zoning devices with adjustable loop response times allow you to use fast-acting devices like glass-breakage detectors, as well as slow-acting devices like contact switches in the protective circuit. Unless the zone locator has a fast-response circuit, you may have to use a pulse stretcher for fast-acting devices.

Protective circuit supervision allows you to supervise the protective loops, even on older alarm systems with control panels that do not offer that feature. Adding a zone annunciator to an existing system often is a good way to increase the system's reliability and provide your customer with better security.

Zone locators that arm automatically when the alarm system is armed require no additional action by the customer. The zone panel is activated automatically by a voltage output from the control panel.

On-site identification of individual zones is useful to the customer and your troubleshooter if service is required. Also, individual zone outputs can tell your central station or monitoring facility which zone triggered an alarm, and police or fire emergency units can be dispatched more efficiently.

Zone-shunting features, either selective or automatic, are worth considering. The ability selectively to shunt individual zones lets both the customer or service technician shut down a problem zone until the problem can be determined and corrected. Automatic zone shunting helps prevent nuisance alarms. With the automatic feature, after the first zone is tripped, the zone annunciator circuitry automatically removes that zone from the protective circuit while leaving the rest active. This feature is particularly useful in stopping "runaway" detectors and frequent swingers from sending numerous nuisance alarms.

Zone annunciators, as add-on devices, may be circuit boards that can be installed inside control panels or separately housed units. Some offer customer-operable features. For example, the zone locator shown in Figure 15.4A allows the customer to bypass individual zones by using slide switches on the front of the panel. The zone locator shown in Figure 15.4B only has visual indicators on its front panel, prohibiting customers from making any adjustments or modifications.

Customer-controllable features can be simple or elaborate. Your customer's requirements will determine the type of zone annunciator best suited for his needs.

If the alarm system is to be monitored by your central station or monitoring facility, you might want to consider installing a zone annunciator that provides individual zone outputs. When an alarm is triggered, you will know its exact location.

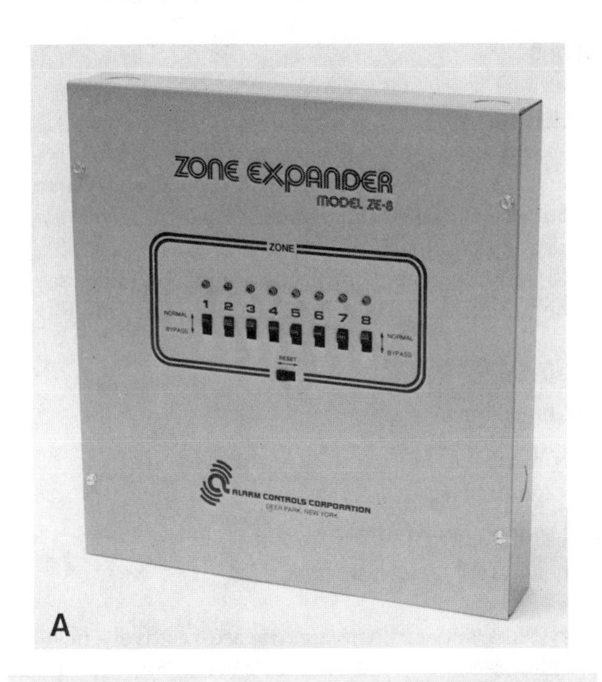

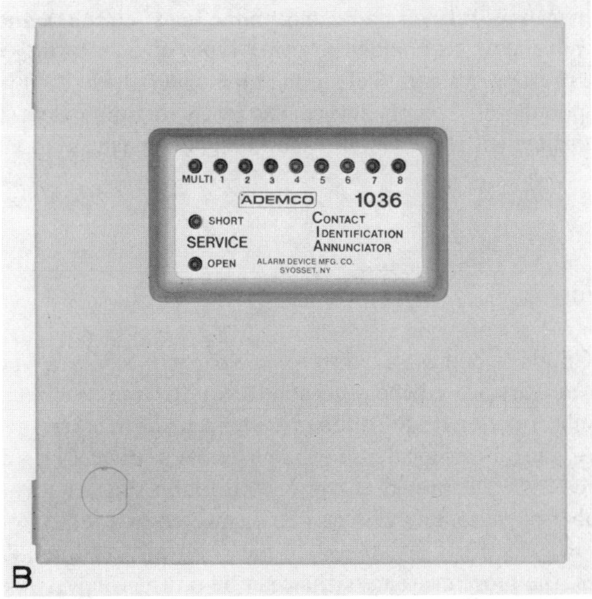

Figure 15.4 **(A)** Zone annunciators with customer-controllable features like individual zone shunting provide your customers with more flexibility in using their systems. (Photo courtesy of Alarm Controls Corp.) **(B)** If you do not want your customer to have access to an annunciator's controls, select a model that provides only visual information on the front panel. (Photo courtesy of ADEMCO.)

REMOTE ANNUNCIATION

Local alarms are good and serve a worthwhile purpose. Alarm systems become much more effective if they are monitored, however. Whether a signal goes to your own central station or to a contract monitoring service, you help ensure that someone will respond to an emergency signal, which may not be the case for only a local bell or siren. Several methods of remote monitoring exist:

McCulloh circuits
Dedicated/direct lines
Multiplex systems
Telephone tape dialers
Digital dialers/communicators
Radio frequency (RF)/radio telemetry

Multiplex systems and radio telemetry are relatively new technologies and are not yet in widespread use, but they are becoming very popular. McCulloh circuits and direct lines rely on dedicated phone lines (circuits). Increasing phone line costs and advancing technology in other types of monitoring equipment have slowed the use of these systems. Telephone tape dialers have been in use for years, but are not as popular as they once were. Currently, digital dialers/communicators are providing dealers with an economical and reliable means to monitor customers' alarm systems.

McCulloh Circuits

Think of a McCulloh circuit as an old telephone party line. Several alarm subscribers are connected to a dedicated metallic conductor (telephone circuit). To identify an individual subscriber on a McCulloh circuit, each alarm is assigned an identification number. A transmitter located on the customer's premises sends coded pulses that are decoded and interpreted at the central monitoring facility.

The number of customers who can be connected to a McCulloh circuit varies depending on the electrical resistance of the transmission line. The shorter the overall distance, the more customers that can be on the loop. A McCulloh circuit is often referred to as a McCulloh loop because the alarms are connected in series, forming a large loop from the central station through customers' premises and back to the central station.

A potential problem for McCulloh loops is "clashing." That is, if several alarms are received simultaneously, the result is a garbled signal that cannot be interpreted at the central station.

Direct Lines/Dedicated Circuits

Direct lines have been used for many years to transmit alarm signals from protected premises to central stations. For certain classes of risks, Underwriters Laboratories (UL) require direct connections for alarm system certification. This type of monitoring circuit provides a direct, dedicated phone line, used only for alarm signaling, between the protected facility and the receiving panel at the central station.

A low level of DC voltage is applied to the telephone line. By monitoring the current flow through a bridge network, the line is supervised.

One cannot always lease telephone lines, so dedicated lines may be difficult to obtain. Moreover, steadily increasing rates for dedicated BA (burglar alarm) circuits may price them out of reach for many potential customers.

Multiplex Systems

Multiplexing allows numerous pieces of data to be transmitted over a single communications channel. Transmission paths can include radio, microwave links, telephone lines, coaxial cable, or simple twisted wire pairs, or signals can be transmitted over a combination of paths. Voice-grade telephone lines probably are the most popular paths.

Telephone Tape Dialers

The often-maligned telephone tape dialer has a place in certain alarm systems. It must be applied and installed properly; moreover, your customer must fully understand its operation and the limitations of its use.

Very simply, a tape dialer sends prerecorded voice messages over voice-grade telephone lines. When the alarm is activated, the dialer seizes the phone line, disconnecting all other phones on the line, dials a prerecorded telephone number, and plays a prerecorded message. Depending on the type of dialer and the length of its tape, several messages may be transmitted.

One drawback to a telephone tape dialer is similar to that of installing only a local alarm: someone must hear the alarm and take appropriate action. If no one is home when the dialer calls, the dialer plays its message anyway. It cannot tell whether or not anyone answered the phone or if the line was busy. To alleviate this problem, some dealers program dialers to call directly into the police department. Before you do this, however, check with them to determine their policies about such calls and whether or not an ordinance governs dialer installations. Many police departments now prohibit such connections.

Even if you do not have a dialer call the police, you can program it to call an answering service or a customer's relative or neighbor. As long as telephone tape dialers are not misused by customers and are applied and installed properly by alarm dealers, they are good annunciation devices.

An interesting use of telephone tape dialers is in conjunction with other remote monitoring techniques. Consider programming a telephone tape dialer to call your customer's radio paging unit ("beeper"). If he usually carries his pager, he will know of an alarm activation as soon as your central station does. You may want to code the transmitted message, for example, "Special alert. Condition three," or something similar. Your customer will know the message's meaning, but no one else will.

Digital Communicators

Digital dialers/communicators also use voice-grade telephone lines to transmit alarm messages. Unlike telephone tape dialers, however, digital dialers send signals to a special digital receiver.

A digital dialer/communicator monitors the status of your customer's alarm system and communicates this information to your central station's receiver. Because its message is digitally encoded, it only can communicate with receivers that can interpret its message.

When an alarm occurs, the communicator seizes the premises' telephone line and dials the telephone number for the receiver. Upon receipt of the call, the receiver acknowledges the presence of the communicator with a special signal ("handshake"), instructing the communicator to transmit its message. After completion of the transmission, the receiver sends another special ("kiss-off") signal, telling the communicator to shut down and reset.

Digital communicators are available as add-on, or slave, units such as the one shown in Figure 15.5, or built into a control panel's circuit board. They are commonly available with three, four, six, and eight channels (outputs).

Digital receivers can handle calls from hundreds and with some models, thousands, of digital communicators. The number of communicators connected to a single receiver should be determined by the types of customers you have and the services you are providing. For example, if your commercial customers report openings and closings by a digital communicator in addition to alarm signals, extra receivers may be needed to handle calls during peak periods, usually mornings and late afternoons.

Although a printer is not a necessity, it makes record keeping easier and more efficient. Some receivers, such as the one shown in Figure 15.6, have built-in printers. Others accept separate printers.

Be sure to consider having a spare receiver in case your primary unit needs repair. If your receiver uses replaceable, plug-in circuit cards, you should have replacement cards on hand. If you are using a printer, a spare is a good idea, too.

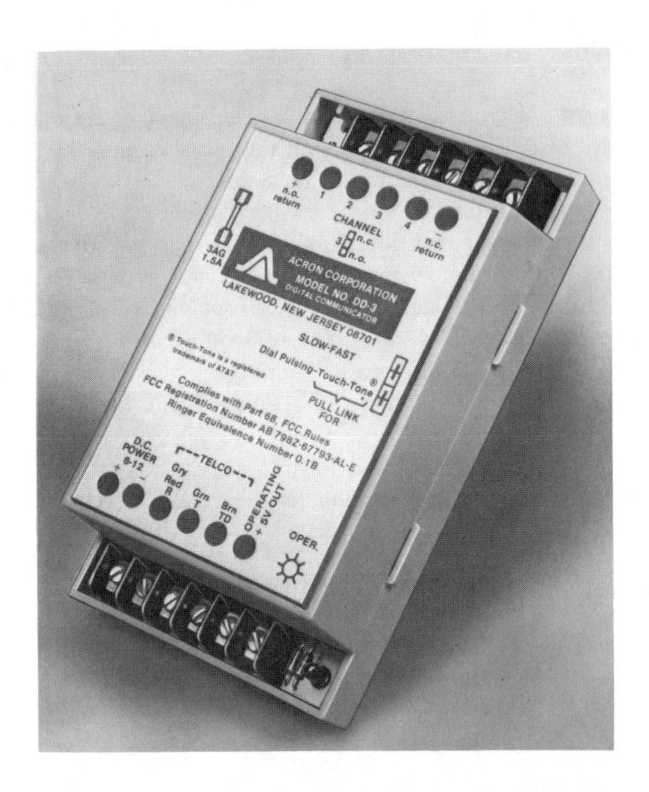

Figure 15.5 Digital communicators are available with a variety of features. For example, the four-channel communicator shown here incorporates automatic low-battery and full restoral reporting for each zone. It can dial by Touch-tone or rotary impulse and be programmed for dial tone detection, second-number dialing, dialing pauses, and reporting delays. (Photo courtesy of Acron Corp.)

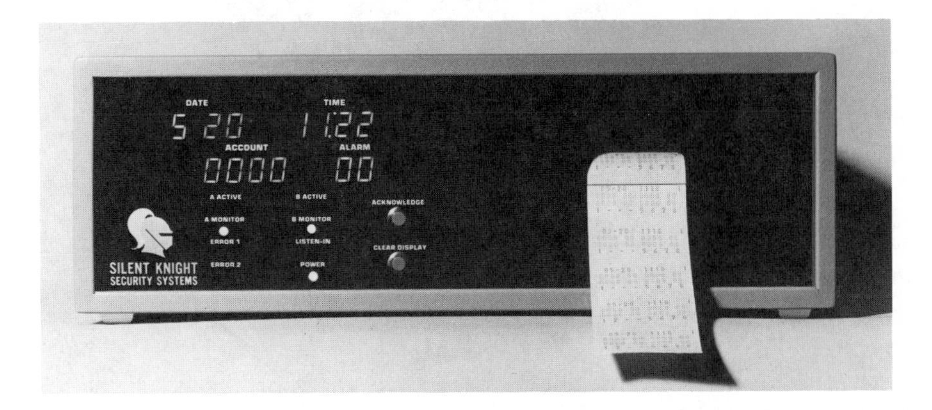

Figure 15.6 Receiver/printer combinations provide visual notification and a permanent printed record of all signals received. (Photo courtesy of Silent Knight Security Systems.)

Start-up Pointers. If you are just starting in business or are planning for your own central station some day, the planning you do now may save you time and money in the future. Consider the following points.

If you will be connecting your subscribers to someone else's receiver, plan for the day you'll have your central station. You might consider purchasing your own receiver and getting your own telephone number, even if you have someone else do the monitoring. That way, when you open your own central station you will not have to reprogram all your customers' communicators.

If you are using a remote contract monitoring service, think about getting your own telephone number with a redial feature. Program your communicators to your local number; the redial feature will automatically transfer calls to your monitoring service. When you are ready to start your own central station operation, you will already have your own phone number (for your receiver) and your communicators will not have to be reprogrammed.

Handy Accessories. Two items will be useful accessories if you install digital communicators. One will help you locate telephone company RJ31X jack problems; the other will provide your customers more security by signaling telephone line problems.

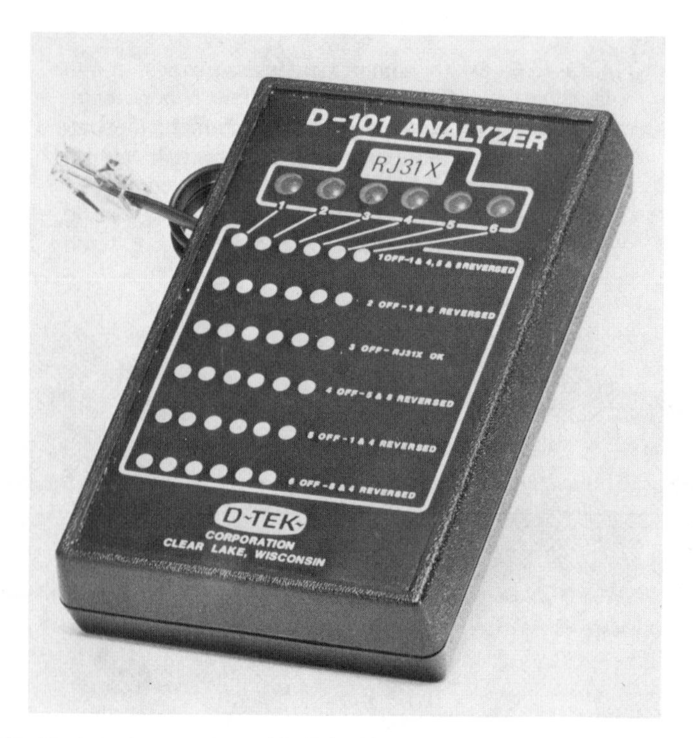

Figure 15.7 Test devices such as this RJ31X analyzer speed up installations by locating and diagnosing telephone line problems. (Photo courtesy of D-Tek.)

Figure 15.8 Telephone line monitors like Adcor's TLM-5 offer an added measure of security. If your customer's telephone line is cut, the TLM-5 can trigger a local beeper or buzzer, or you can trip a twenty-four-hour protective circuit on his alarm system.

A telephone circuit analyzer, such as the one shown in Figure 15.7, plugs into an RJ31X jack and automatically diagnoses problems. It can tell if wires are reversed or if the circuit is good.

A telephone line monitor helps increase the customer's security by triggering an alarm if the telephone line is tampered with or cut. The device shown in Figure 15.8 attached directly to the telephone line or digital communicator. If the telephone line is cut, the line monitor can be connected to trigger an immediate local alarm.

Radio Telemetry

Radio telemetry, or long-range wireless systems, can be used to transmit an alarm from a protected premises to your central station. They monitor systems in areas where telephone lines do not exist or may be difficult to obtain. Some dealers use radio telemetry in addition to other signaling methods to provide redundancy.

Radio frequency (RF) systems are usually line-of-sight. That is, transmitters and receivers must be in view of each other. Any large objects, like hills or buildings, between the transmitter and receiver will block and scatter the RF signals.

Radio telemetry is becoming increasingly popular with alarm dealers. With the deregulation of American Telephone and Telegraph and the uncertainty of telephone line availability and rates, RF alarm signal transmissions may be a viable alternative. It is definitely worth watching.

MONITORING BY OTHERS

You can provide your customers monitoring without having your own central station. Remote monitoring centers throughout the country offer their services at competitive prices. For only a few dollars a month, your customers can choose from a wide range of services.

National monitoring centers provide WATS lines for your communicators to signal their receivers. Trained operators respond to the signals and notify the appropriate authorities according to your instructions. Most of these monitoring centers are highly computerized.

Many offer special services to dealers. They can assist you with advertising and marketing information or customize their services to meet special needs. For example, Emergency 24, a Chicago-based national monitoring center, offers an emergency information/identification card to its dealers' subscribers. The card, called Life Card, has a toll-free telephone number printed on the back. In an emergency, paramedics, police, or others can call the telephone number and give a special code to the Emergency 24 operator. The operator will take the information and contact the cardholder's relatives or friends based on the data stored in their computer.

Initially, remote monitoring will provide you with the services you need to be competitive in the alarm industry. Whether you decide to continue with a monitoring service or start your own central station will depend on your objectives. You can build a good, recurring monthly income from monitoring fees in the meantime. Generating ten dollars per month from a customer's monitoring account may not seem like much, but multiply that amount by 100 or 1,000 customers and you can see why recurring revenue should be a part of your business planning.

Specialized Equipment 16

Although the equipment and systems evaluated in this chapter could have been discussed in previous chapters, they have some unique characteristics and should be considered separate from other items.

Wireless alarms employing radio frequency (RF), ultrasound, and line-carrier technologies are becoming increasingly popular. Ultrasound and RF systems now offer supervision to tell you the status of each transmitter or transponder in the system. Line-carrier systems will probably have similar supervisory features in the near future. In this chapter, we discuss the new, high-tech supervised RF and ultrasound systems.

Hold-up alarms are usually simple, but proper application and installation of equipment is critical because these systems must work in life-threatening situations. Whether a hold-up alarm is part of an intrusion system or a stand-alone system, it provides additional marketing opportunities for aggressive dealers.

In many respects, medical alert systems are similar to hold-up alarms. Both use much the same type of equipment and both respond to the needs of a specialized market.

Outdoor protection systems can be a dealer's nightmare if not installed properly. If you can master their installation, you will find them profitable additions to the other security products and services you offer.

All of these specialized systems offer you an opportunity to open new markets or become more competitive in your present markets. If you are looking for avenues of growth, don't overlook these.

WIRELESS ALARMS

Wireless alarm systems have been used for many years. Recently, however, technologic advancements have provided alarm dealers with supervised wireless equipment that can check on itself and report its condition to the alarm-control unit.

Supervised wireless has overcome the two common complaints about nonsupervised units: battery condition and circuit status. Until now, you could not tell if a transmitter's batteries were low unless you physically tested each one. Also, you did not know the status of the protective circuit: you could have one or more doors or windows open and still arm the system, leaving areas unprotected.

In addition to checking their own batteries and telling you circuit status, transmitters can provide other information. For example, they can periodically "check-in" with the receiver, just to tell the receiver that everything is working normally.

Although transmitters can be connected to most protective devices (Figure 16.1), some detectors seem ideally suited to having transmitters built into their components. Currently, some passive infrared detectors, audio switches, and smoke detectors are available in wireless versions. Many of these stand-alone wireless devices are powered by internal batteries, simplifying installations and reducing labor costs. Figure 16.2 shows a small, attractive, passive infrared detector (PIR) with a built-in wireless transmitter.

Few systems are totally wireless. Wires are needed to connect keypads and

Figure 16.1 The ultrasonic transducer is connected to two magnetic contact switches. Instead of using RF wireless, it uses ultrasound to transmit alarm and status information. (Photo courtesy of Ultrak, Inc.)

power supplies to control panels, and speakers to siren drivers and detection devices to transmitters. For all practical purposes, however, wireless protection is here.

If you are wondering if wireless systems are suitable for your company's operation, use the worksheet that follows to compare hard-wired versus wireless alarm systems.

"IS WIRELESS COST EFFECTIVE FOR YOU?"[1]

Proponents of wireless alarm systems claim that installation costs are reduced because of labor savings. Opponents say that those savings are offset because wireless equipment costs more.

---

[1] "Is Wireless Cost Effective for You?" Reprinted by permission from *Security Distributing & Marketing*. © February 1983, Cahners Publishing Co.

Figure 16.2 Because passive infrared (PIR) detectors are small and require little power, they are very compatible with RF transmitters. Units such as the one shown have their own internal power supplies that operate both the PIRs and transmitters. (Photo courtesy of Colorado Electro Optics.)

The worksheet shown below will allow you to compare wireless and hardwired installations to determine which is most cost effective for *you*—based on your individual operation.

Major variables, such as labor rates, installation time and additional equipment, have been included in the worksheet. One variable not included is the time to install the receiver and/or interface. You may want to consider that variable before making a final determination about cost effectiveness.

Step 1.	Enter installer's hourly wage:	$_____	(1)
Step 2.	Multiply (1) by 1.267* for loaded labor rate. Enter amount here:	$_____	(2)
Step 3.	Enter cost of wireless receiver:	$_____	(3)
Step 4.	Enter cost of wireless interface, if used:	$_____	(4)
Step 5.	Add (3) and (4) for total receiving equipment cost. Enter amount here:	$_____	(5)
Step 6.	Enter number of transmitters in system:	_____	(6)
Step 7.	Divide (5) by (6) for proportional cost. Enter amount here:	$_____	(7)

* 1.267 is the estimated loading factor, including employer's contribution to FICA (0.067), benefit package (0.15), and miscellaneous payroll taxes and costs (0.05) added to the basic labor rate (1.0).

Step 8.	Enter cost of one transmitter:	$_____	(8)
Step 9.	Add (7) and (8) for loaded transmitter cost. Enter amount here:	$_____	(9)
Step 10.	Enter cost of detection device:	$_____	(10)
Step 11.	Add (9) and (10) for total protection unit equipment cost. Enter amount here:	$_____	(11)
Step 12.	Enter estimated time to install one transmitter and detection device:†	_____	(12)
Step 13.	Multiply (12) by (2) for protection unit installation cost. Enter amount here:	$_____	(13)
Step 14.	Add (11) and (13) for total wireless protection unit installed cost. Enter amount here:	$_____	(14)
Step 15.	Multiply (14) by 0.25 for adjusted cost.‡ Enter amount here:	$_____	(15)
Step 16.	Enter estimated time to install and *hard-wire* detection device:†	_____	(16)
Step 17.	Multiply (16) by (2) for hard-wired protection unit installation cost. Enter amount here:	$_____	(17)
Step 18.	Add (10) and (17) for total hard-wired protection unit installed cost. Enter amount here:	$_____	(18)

The Bottom Line

The calculations in the worksheet provide two figures for comparison: a wireless installation cost (15) and a hard-wired installation cost (18).

Cost comparison: $________(15) vs $__________(18)
Wireless Hard-wired

For example, 10 minutes = 0.17 hours and 15 minutes = 0.25 hours. (See Page 54 of *SDM,* September 1982 or page 62, October 1981, for suggested installation times.) Author's note: See Table 10.1 in Chapter 10 for suggested installation times.

HOLD-UP ALARMS

As I mentioned at the beginning of this chapter, hold-up alarms are relatively simple to design and install. They require only a few components: a switch, a power supply, and an annunciation device.

---

† Convert minutes to fractional hours using decimals.

‡ The productivity factor of 0.25 means that four wireless systems can be installed for each hard-wired system. (If your productivity factor varies, substitute the correct number.)

Digital communicators are good annunciation devices. They operate silently so they will not startle the robber, but send an emergency message to a monitoring center. Police officers respond to hold-up alarms quickly, realizing the potential for danger when they arrive at the scene. It is not unusual for several squads to arrive with officers carrying shotguns. It is imperative, therefore, that hold-up alarms be designed so that they are not triggered inadvertently.

Whether this alarm is a separate system or part of an intrusion alarm system, it should be ready to report a hold-up at all times. Your customer should never be able to turn it off. It should be installed so that it can be actuated with a minimum chance of detection. Cash drawer money clips are often good for this purpose, however, they are among the easiest devices to trigger accidentally. Removing the last bill in the cash drawer, the one inserted in the money clip, triggers the alarm, so care must be exercised when removing the money at the end of the business day. Money clips are available in wireless and hard-wired versions.

Hold-up buttons provide another means of tripping the alarm. Most are designed to prevent accidental triggering. They have either two buttons to press simultaneously or one recessed button.

Portable (wireless) hold-up buttons are often useful and can be carried in a pocket or purse, or on a belt clip. Be sure that your customer understands the range of the wireless device so he will not attempt to use it from too great a distance.

If hand actions might be observed by a would-be robber, consider using a foot-rail switch that is triggered by a simple foot movement. Here, too, care must be exercised and customers educated, so the device will not be tripped accidentally by an employee or a cleaning person sweeping the floor.

Hold-up devices should be installed at all locations where a robbery might occur, especially, near cash registers and other money-handling areas. Moreover, consider installing back-up devices in other locations such as storage rooms, rest rooms, and vaults in case employees are locked in these areas. You might want to consider installing additional hold-up buttons at the receptionist's desk, switchboard, and in the manager's office.

If you are installing a hold-up alarm system in a large facility and will be using numerous buttons, foot rails, and/or money clips, consider zoning the system. It will help your central station operator dispatch the police to the appropriate location with minimum delay.

Residential hold-up alarms, usually called "panic" systems, provide an opportunity for additional sales with a home security system. Opinions differ about whether a panic alarm should be silent and report to a monitoring center, or loud and sound a bell or siren.

Why not do both? Using a two-channel wireless RF transmitter you can provide two levels of panic protection. For example, pressing the left button on a hand-held transmitter could trigger a local alarm; pressing the right button could trigger a silent alarm to your monitoring center. Pushing both buttons, sequentially, not simultaneously, would give your customer a loud alarm and police notification.

MEDICAL ALERT SYSTEMS

A residential panic alarm system by another name is the medical alert system. Instead of notifying the police, your central station notifies paramedics or others. The components for the two systems are identical.

Additional features make medical alert systems unique and more useful, however. An important feature to consider is daily check-in. If a person (patient) fails to check in once every twenty-four hours, an emergency notification is sent to the monitoring facility, a feature that is useful to elderly persons who are shut-ins. Pressing the check-in button keeps them in touch with the outside world; failing to do so once every twenty-four hours summons help.

Another system uses a two-channel wireless RF panic button. Pressing one of the buttons is the same as pressing the emergency button on the front of the console: a medical emergency message is sent to a monitoring facility by a digital communicator. The second button on the transmitter can be used to send a panic or other signal.

Installation of medical emergency systems is simple. After the digital communicator is programmed and the codes are set on the wireless RF receiver, all you have to do is connect the unit's transformer to an AC outlet and plug its telephone cord into a phone jack.

You may need to spend extra time educating customers about using medical alert systems. They are not complicated, but you want to be sure that inadvertent alarms are minimized.

Theoretically, a patient with a medical alert system prescribed by a doctor would be able to deduct it from income taxes as a medical expense. Check with your accountant or tax advisor for the latest Internal Revenue Service rulings. Offering it on a sale or lease basis in conjunction with tax advantages may help you market these systems.

OUTDOOR SECURITY

Outdoor security is challenging. It also is rewarding if done properly. Otherwise, it is a series of headaches.

Any facility that stores material outdoors, is plagued by vandalism, or needs to know at the earliest possible moment when a potential intruder is near, is a prospect for an outdoor security system. These types of prospects include schools, lumber yards, utility companies, trucking companies and docks, telephone companies, and government installations.

Equipment designed for outdoor use is made to withstand tough environmental conditions. Even so, it must be installed properly and within the limits of its specifications. It must perform through scorching summers and sub-zero winters.

The advantage to an outdoor security system is that it detects intruders before they have an opportunity to reach a building. Police and/or guards can

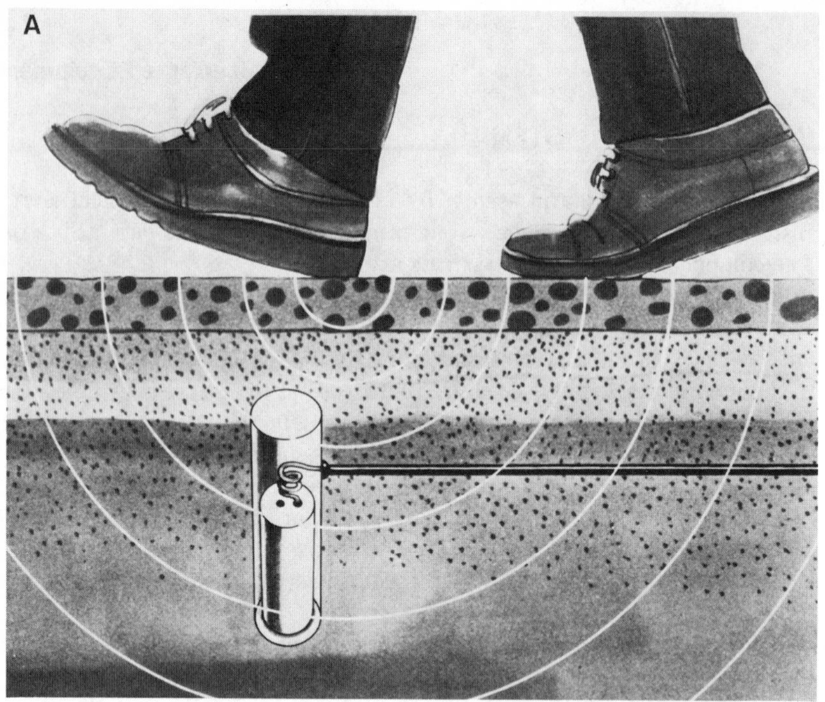

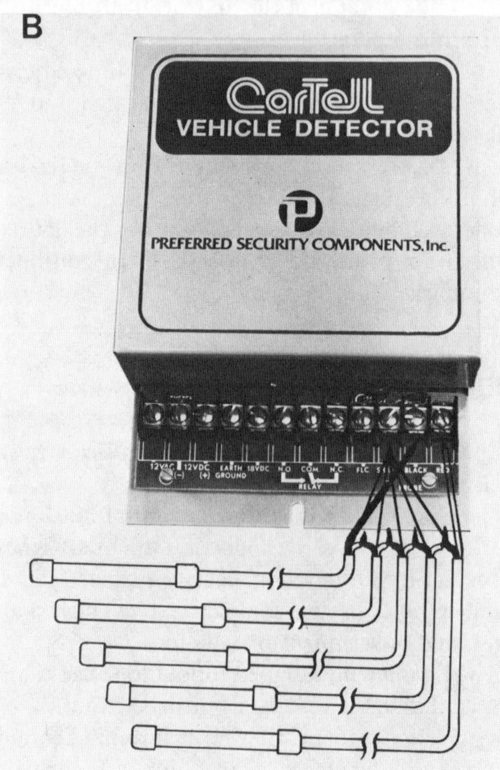

Figure 16.3. **(A)** Buried sensors react to seismic waves and pressure to detect intruders. (Photo courtesy of Van der Heide Totale Preventie and the Consulate General of the Netherlands.) **(B)** Buried vehicle detectors respond to metallic mass. Large, iron-based objects will trigger an alarm. (Photo courtesy of Preferred Security Components, Inc.)

respond at the earliest possible moment to prevent further entry. Devices are generally in one of three categories: buried, fence, or space sensors. The types of detection devices you use will be determined by your customer's needs and the local terrain.

Buried sensors, such as the one shown in Figure 16.3A, can form a protective zone along the ground under which they are buried. Intruders walking across the ground create seismic waves and pressure that are detected by the sensors. Others, like those in Figure 16.3B, respond to magnetic stimuli. Iron-based objects, such as vehicles, passing over or near the sensors will trigger an alarm.

Fence-disturbance sensors are known by a variety of brand names and detect intruders climbing over, crawling under, or cutting through chain-link and other fences. Usually, these sensors respond to vibrations caused by an intruder as he moves the fence fabric.

Fences can be protected in other ways, too. Electric-field sensors, composed of parallel field and sense wires, generate electrostatic field. An intruder's body disturbs and distorts the electrostatic field as he approaches the fence. Gates can be wired with wide-gap magnetic contact switches. The switches should be especially designed for outdoor use, and armored cable is recommended for wire runs. Outdoor infrared photoelectric devices work like indoor units: breaking the beam triggers an alarm. Outdoor microwave units operate differently from indoor units. Instead of detecting movement, as indoor units do, outdoor units detect either a reduction or a distortion in microwave energy between the transmitter and receiver. Infrared beams are narrow and microwave beams are broad. Both have special uses and applications.

Because infrared and microwave units are line-of-sight devices, anything blocking their signals may cause an alarm. It is important, therefore, that areas between transmitters and receivers be cleared of debris, obstructions, and vegetation.

Snow, rain, and fog will affect outdoor space protection devices depending on how much and how rapidly they obscure and the types of detection devices used. Many of the newer outdoor units have compensation circuits to prevent inadvertent alarms if obscurity is gradual.

Only rarely does a single type of detection device provide good protection for an entire perimeter area, especially if it is large. Usually, several types must be used to afford adequate protection. Often, a secondary, or back-up, system must be used in high-security applications.

Outdoor security systems have potential problems that are truly unique. Animals, like cows, deer, or dogs, may trigger an alarm by running over a protected area or rubbing against a protected fence. Large birds may fly between a space protection device's transmitters and receivers and cause an alarm. You may want to check adjacent property to find out what kind of animals are present and what their movement patterns are. The more information you have about the surrounding area, the better equipped you'll be to design a reliable outdoor security system.

Unique Applications 17

Webster's New World Dictionary, Second College Edition, defines unique as: "1. one and only; single; sole . . . 2. having no like or equal; unparalleled . . ." By strict definition, therefore, most alarm system installations are unique because slight differences exist in almost every one. Many are similar, but it is doubtful that any two are exactly the same.

It is not the purpose of this chapter to provide you with a definitive listing of unique applications—that would be impossible. Rather, the intention is to plant a seed and let you cultivate it. The following examples show how common alarm system technology can be used in different ways to solve less-than-common and sometimes, nonsecurity problems. Applications for alarm system components and systems are limited only by your imagination.

OPEN HOUSE ALERT

One of my favorite unique applications is an open house alert. You can use an automobile alarm system with a radio paging transmitter to notify a real estate agent when someone has entered an open house.

Connect the auto alarm to a regulated 12-VDC power supply and a passive infrared (PIR) detector. Place the PIR near the front door so it will detect prospective home buyers as they enter the house. When the system is triggered, the real estate agent is notified by the paging unit he carries. He then can go to the open house and meet the prospects.

This type of system can be used to allow an agent to show two houses simultaneously or to notify an agent in a nearby sales office that a prospective buyer is on the scene. Although the auto alarm's paging system has a range of several miles, it probably would not be practical for the open houses to be separated by more than a few blocks.

If you want to be more creative with an open house alert system, add a tape player. When prospects enter the house, the tape player could announce, "Please take a few minutes to look around. A sales agent is on his way to answer any questions you might have." This announcement also would serve as a warning to a would-be intruder that his presence is known.

MOBILE PROTECTION

If we can use vehicle alarms in homes, why not use home alarms in vehicles? In lieu of or in addition to vehicle alarms, consider using residential alarm systems in vans used as mobile showrooms. Mobile showrooms may carry thousands or tens of thousands of dollars worth of inventory. You might want to use a zoned alarm system with perimeter and interior circuits.

The perimeter zones could protect doors with magnetic contact switches and glass with audio discriminators. One door might be selected for an entry/exit zone and the other protective devices would be placed on instant circuits.

The vehicle's interior could be protected with space protection devices such as PIRs or ultrasonic motion detectors. Depending on the type of system you install, you could have all control components locked in a cabinet and the entire system controlled by a digital keypad.

FIRE-DETECTION SYSTEMS

When you think of vehicles, don't think only of cars, vans, and trucks, and don't think only of intrusion detection devices.

A Long Beach, California, alarm dealer installed a forty-zone commercial fire alarm system in a tugboat at an installed price of $17,000. Why was such an elaborate fire alarm system needed? Because the tug carried explosives. Detecting a fire at the earliest possible moment is critical.

PORTABLE PROTECTION

Stamp and coin dealers frequently attend bourses where they exhibit and sell their wares. Often such exhibitions last several days. Many of these dealers keep their inventory in locking steel cabinets during the shows. Although security guards are present when the exhibitions are closed for the night, some dealers like to take an extra measure of precaution and have alarm systems installed on their inventory cabinets.

Usually these systems are very simple, key- or keypad-controlled, and relatively inexpensive considering the value of the inventory they are protecting. A siren or horn alerts the security guard to any tampering with the cabinets.

Of course, additional protection could be provided with an alarm system employing a noise maker and a digital communicator connected to an alarm-monitoring facility or your central station. In addition to notifying the security guard, the dealer and/or the police also could be alerted.

DETECTING WALK-AWAYS

A unique application for supervised radio frequency (RF) wireless alarm systems is to detect walk-aways. A supervised transmitter periodically checks in with its receivers. If a check-in signal is not received, a trouble alarm is sounded indicating that a transmitter may be out of range. That is, the person has wandered too far from the central security location. Attempts to tamper with or remove the transmitter signal an alarm, too. Because the transmitters can be individually identified, so can the persons wearing them.

Similar systems can be used in nursing homes and health care facilities. Walk-aways are detected before they wander too far or become injured.

These are five examples of an almost limitless number of unique applications for alarm systems. Expand your horizons. Employ some creativity when solving problems. Numerous possibilities exist for using alarm system technology.

POTPOURRI IV

Computers and the Alarm Dealer 18

Chances are you will need at least one computer as an alarm dealer. If you have, or are planning to start, a central station, a computer is a necessity. Moreover, an office computer should be considered to help you with estimating, bidding, marketing, and accounting. It is unlikely that one instrument can perform both as a central station and an office computer and perform efficiently.

CENTRAL STATION COMPUTING

Automating alarm-signal processing in a central station can be costly and complex. Computer hardware and software can cost thousands of dollars. In some cases, total costs can easily exceed $10,000.

A good system is cost effective and may pay for itself in labor savings. Opening, closing, and other routine supervisory signals can be processed automatically without an operator's intervention. Alarm signals can be processed more quickly and reliably because the possibility for human error is reduced.

Selecting a Central-station Computer

A perfect central-station computer does not exist. You must determine what functions you want a computer to perform, then find one that suits your needs. If you already have a central station and are considering a computer, ask yourself these questions:

1. How many monitored accounts do I have?
2. How many alarm signals are received per day?
3. How many maintenance and supervisory (e.g., opening and closing reports) signals are received daily?
4. How many new accounts are added monthly?

If you do not have a central station but are considering opening one, ask yourself these same questions. You will have to use estimates because you will not have other data available.

The answers you arrive at indicate the volume of signals that your central station receives. The more signals you receive, the more likely it will be that you can justify a computer. It will allow you to increase the volume of signals received without a corresponding increase in your central station staff.

You may want your computer to perform additional functions; for example, maintain customers' account histories, track false alarms, and generate billing documents.

Supplier Considerations

Unless you are a computer programmer and systems engineer, you probably will be better off leaving programming and installation to experts. Several alarm equipment manufacturers offer central-station computers. They have the necessary expertise and industry background to assist you.

You will probably have a continuing relationship with the company that sells you the computer. You should attempt to determine that it is reputable and will be in business at least as long as you will.

Of course, computers can be purchased from sources outside the alarm industry. While these persons may be knowledgeable in computer hardware and software, they may not know alarm company operations. If you use an outside source, work closely with the person who will develop your system's software. Off-the-shelf, or "canned," software is not yet available for central station operations.

System Components

A central-station system consists of a minicomputer or microcomputer, a video display terminal (VDT), a printer, and the necessary interfaces to connect the components. You will also need a digital receiver to receive signals from alarm systems' digital communicators. A printer connected to the receiver is recommended. If your computer system should become inoperable, you can process signals through your receiver and printer manually. If you intend your system to replace your digital receiver and printer, you will need a complete back-up computer system in case your primary system fails.

The computer system's "brain" is its central processing unit (CPU). Data from the digital receiver are fed into the CPU through an interface. They are then displayed on the VDT and/or computer's printer.

It is important to remember that the CPU's memory is volatile. That is, if power fails, all information currently in memory will be lost. A standby power supply for your computer system is not a luxury, it is a necessity.

The data generated by your computer are stored in a file storage device such as a floppy or hard disk. The information stored on disks is nonvolatile; it is maintained even during power outages. Back-up copies of all disks are recommended, however.

Your operational characteristics will determine whether a floppy or hard disk system is appropriate for you. As a general rule, you probably should consider a hard-disk system if your central station monitors more than 500 alarm systems or receives more than 100 signals per day.

The VDT is the interface between your central station operator and your computer system. It displays data generated by the computer and by the operator by a keyboard.

The computer's printer serves the same basic purpose as the digital receiver's printer: it provides hard copy of all events. Of course, because the computer can generate significantly greater quantities of data, more information can be printed.

Software, the programs that make your computer work, is the heart of your system. Without it, the hardware is useless.

Summary of Considerations

A well-designed computer system should make your central station operate more efficiently and cost effectively. It should minimize human errors and almost eliminate missed signals. Because the computer makes certain decisions based on programmed variables, it should be more reliable than manual systems. A good system prompts central station operators through the necessary steps for processing alarm, maintenance, and supervisory signals. The smaller the margin for human error, the better the system.

The computer's purpose is to augment existing manual receiving equipment, not replace it. Digital receivers must be capable of functioning alone in case the computer fails.

A complete printout of all activity helps protect you and your subscriber. Everything that happens through the computer system should be recorded and saved.

Design your system with the future in mind. Your computer should easily accommodate your company's growth.

A MICROCOMPUTER IN THE OFFICE

Together with the rest of society, alarm dealers are witnessing a revolution in computer technology. Microcomputers offer computing capabilities previously reserved for large systems. For years, only large alarm companies used computers. Smaller companies now are discovering the potential of microcomputing technology.

A microcomputer offers several advantages over large mainframe computers. First, of course, is cost. Microcomputers are considerably less expensive than mainframe computers. Second, they are user friendly. Often, they can be programmed without the services of a computer programmer. Third, they are interactive, that is, the user knows immediately the results of his actions.

Computer Capabilities

Before you purchase a microcomputer, consider its capabilities and limitations. Regardless of what a salesperson tells you, you will not be doing wonderful things with your new system within two hours after plugging it in. Chances are you will not have completed reading the instruction manual in two hours.

Computers are only as smart as their software and only as accurate as the data put into them. Computer operators have a term for data quality: GIGO—garbage in, garbage out. If you put bad numbers into your computer, you will get bad numbers out of it, but you will get them out very fast.

Sometimes a computer is nothing more than a very fast idiot. It will turn out reams of data for you. The question is: how many of the data are useful? The important consideration is the quality of the output, not the quantity. Computers are useful tools if they are used properly.

Selecting Software

Before you even think about hardware, think about software. The software you need may only run on certain machines. If so, your hardware choices will be limited. Several general types of microcomputer software presently are popular.

Electronic spreadsheet programs allow you to compare mathematical relationships and make calculations based on these relationships. For example, you can project sales and profits based on estimated growth patterns.

Filing or data base programs allow you to enter, receive, sort, and index data quickly. This type of program is especially useful for tracking equipment or installation problems and false alarms.

Word-processing programs allow you to compose text on the video display screen and print it out after corrections have been made. The text may be saved on disks or tapes for later use. Word-processing programs help simplify creating forms and form letters.

Statistical processing programs allow data to be processed for various statistical functions, including standard deviation and histogram analysis. These programs are helpful in describing the characteristics of data or in analyzing them for correlation between various factors.

Graphics programs can create pie charts, line charts, and histograms from data and print these graphs on paper.

Communications programs allow you access to various information services by way of telephone connections. They allow data to be transmitted between computers over telephone lines. They are popular because you can tailor them to fit your needs and because they perform useful functions with little programming knowledge. You might consider them "generic" programs; they are good, but they might not meet some of your specific needs. Microcomputer programs also are available for specific functions like general ledger or accounts receivable accounting.

Shopping: Choices and Options

Whether you are shopping for your first computer or are purchasing a second or third, you'll be faced with some decisions. Here are some things to keep in mind when choosing a personal business microcomputer.

Memory. There are two basic rules regarding a computer's memory. First, larger memories can make complex programming more efficient and allow you to do more sophisticated calculations. Second, the larger a computer's memory, the higher the price.

Simply stated, memory is a computer's warehouse where instructions and data are stored. The warehouse is divided into electronic storage bins called locations or addresses. Each address can store one byte of information (one byte equals eight bits—or *bi*nary digi*ts*).

A byte can contain a single alphanumeric or graphic character, part of a number, part of an address for another memory location, or a single instruction for the processor. A byte is a very small parcel of information. Many memory locations will be needed, therefore. Microcomputers generally are limited to 65,536 locations. This size of system usually is referred to as a 64k system. The k stands for kilo, or thousand. Although the computer's memory exceeds 64k-bytes (kilobytes), it is still referred to as a 64k system. Some microcomputers can access more than 64k-bytes, but we will treat that as our ceiling for our purposes here.

To be programmable and also perform routine chores, like displaying information on the screen and saving and loading programs, two types of memory are required: ROM and RAM. Both types reside in the 64k-bytes mentioned above.

*R*ead *o*nly *m*emory (ROM) is for permanent storage; *r*andom *a*ccess *m*emory (RAM) is temporary storage. Both are random access memories, but ROM is written by the computer manufacturer and cannot be changed by the user.

To understand the difference between ROM and RAM, look at it this way: ROM is like a slab of granite with information chiseled deep into its surface; RAM is like a chalkboard. The contents of its memory can be written, read, erased, and rewritten, an operation that can take place in a few millionths of a second.

The size of ROM is not as important as the size of RAM, because that is what stores your programs and data. Larger RAM allows larger quantities of data to be entered and processed.

Some programs require large memories. For home use, 16 or 32k may be sufficient. For business use, you will be better off with 64k. Newer microcomputers, with proportionately higher price tags, offer 96k, 128k, and 256k memories.

Screen Displays and Formats. Your display is your window into your computer. Without a screen, typing into your computer is like using a typewriter with no paper.

Some microcomputers allow the use of standard black-and-white or color television sets. If you are purchasing a computer with color graphics, you'll need a color television or color video display monitor. While your computer probably will operate with either of these, chances are that the monitor will give you better service. Connecting your computer to the cheapest television you can find is somewhat like purchasing a $2,000 stereo and hooking it up to $9.95 speakers. Video quality counts as much as sound quality counts on a stereo. Some computers have built-in display monitors. That is another point for you to consider.

Common computer screen formats offer 22, 24, 32, 40, 64, and 80 columns. The most versatile format for business use is 80 columns. Also, your computer should display at least 16 lines. Your best choices would be a 24- or 25-line format. The larger the format, that is, the more columns and lines, the more information you can display at one time. Larger formats are easier to read, too. They are more expensive as well. You'll have to balance cost against need.

Keyboard. A keyboard is not really part of the computer. It is an input device, a "peripheral." Most computers come with keyboards, so it is not something you'll have to shop for. Do not underestimate its importance, though. If you, or one of your staff, are to spend hours at the keyboard, you will want to be sure that it is comfortable and easy to use. A keyboard's human engineering is critical.

Software. The software that you need to do the job you have in mind may determine the hardware that you eventually purchase. Numerous magazines and books are available that discuss and evaluate software.

Your first task before you write out a check for a computer is to determine software availability. Do your homework and you'll make a wise purchase. Otherwise you may find that your computer system won't do what you thought it would.

Peripherals. Will you need any peripheral equipment? Almost certainly you will if you want your system to perform to its full capabilities.

You already have two peripherals: the keyboard and video monitor. The other primary peripheral you may want is a printer. Without a printer, your use of your system will be limited to the data displayed on the monitor.

If your computer will be communicating with other computers, you'll need a modem, short for *mo*dulator/*dem*odulator, which connects your system to the telephone network.

Storing Data. You must have a method for storing programs and data. Two major types of storage are available: tapes and disks. The major differences between them are cost and speed. Disk drives add several hundred dollars to your computer system's price tag, but they allow you to load, save, and retrieve data quickly. They also can store much greater quantities of information than tapes.

Documentation. The documentation, instruction, and programming manuals provided with your computer are as important as those that come with the alarm components you buy. Good documentation is essential. Your understanding and the ultimate usefulness of your computer depend on the quality of the documentation. Fortunately, for those computers having poor documentation, someone usually writes a book that you can purchase to help you with your system.

Spending several thousand dollars for an office microcomputer system is an investment you should make wisely. Careful planning now will help ensure that your investment pays future dividends.

The Legal Side 19

We live in a free country. Because of that, you can become an alarm dealer after meeting only minimal requirements. You can sell and install security products and provide related services, and collect money for doing so. As long as your company remains small and your income is reported honestly, you'll have few legal requirements to worry about. Or so it would seem.

What would happen if one of your customers suffered a loss? Like it or not, by being in the alarm business, you have assumed certain liabilities. Of course, general liabilities are created with any type of business. Specific liabilities are created because you, an alarm dealer, have agreed to help protect customers' lives and property.

Few regulations exist making liability insurance mandatory. That does not mean that you do not need any, however. You and your company should be adequately covered with insurance because the law may find you liable for damages or losses that occur. To begin a business you must provide some safeguards both for the public and for yourself. As your business grows from a one-man operation and you hire people to work for you, you will have additional legal requirements.

Taking in a partner or starting a corporation imposes legal obligations on you and your company. At some point during your company's early growth, you may find these becoming burdensome. The paperwork involved may take the fun out of owning your own business.

INVALUABLE ADVISOR

An attorney may well be the most important advisor your company can have. It is important to start a good working relationship early, as a lawyer who has been with you from the beginning will be better able to help you as things get more complicated.

Attorneys' services can be expensive; the lack of their services can be even more expensive. Do not look at attorneys' fees as expenses. They are investments in your company's future.

Periodically, you and your attorney should review your operations:

- To protect your personal assets
- To protect your business' assets
- To protect partners in a partnership or principals in a corporation
- To determine which tax category is most beneficial
- To protect your successors or heirs
- To make sure you comply with municipal, county, state, and federal laws
- To help you prevent problems with unions, discrimination, licensing, zoning, and the like

Legal Complexities

Because you deal with the public, you may someday be involved in a legal action. Without an attorney, you will find yourself alone, attempting to defend yourself and explain the complexities of electronic security systems to plaintiffs, lawyers, and a judicial system.

You may be 100% certain that you, your employees, and the products and services that you provided were not responsible for a loss. Convincing a jury, judge, or insurance company adjuster may not be easy, however. Your attorney should be at least conversant with electronic security terminology.

Contractual Defense. In a legal situation, your contract is your best defense. A well-written contract protects you and your customer.

An all-purpose, fill-in-the-blanks contract for alarm installations is not worth the paper it is written on. I have seen alarm dealers use these so-called contracts, which are nothing more than sales orders. I shudder when I think of what might happen if these dealers are ever sued. Unfortunately, I suspect that they'll have no defense against claims regardless of how well the alarm system worked or was installed.

Your contract must be designed to meet legal requirements in your market area. A contract that works in Maine may not work in California. Because specific contract provisions vary from locale to locale, it would be futile to list them here. Some general provisions that should be contained in every alarm installation contract include:

1. Disclaimer of warranty: a burglar or fire alarm system will not prevent a burglary or fire. (No such claims should ever be made.) An alarm company cannot guarantee that the service it provides will be effective.
2. Noninsurer: your alarm company is not an insurance company. It may be obvious to you that your company is not insuring against a loss. It may not be obvious to a customer, however, and a noninsurer statement should be included in the contract. You are charging them for a security system based on the system's value, not on the protected property's value.
3. Limited liability: your contract should limit your company's liability even if you or your employees were negligent. A specific amount, usually $250, is stated in the contract; thus your maximum loss would be that amount.
4. Third-party claims: you also must be protected against third-party claims. This clause means that persons other than your customer cannot make claims against you. For example, if your customer is a furrier and is burglarized, his customers (who may have furs stored in his vault) cannot file claims against you.

Your contract also should specify, in detail, what services you are providing and what responsibilities belong to your customer. These provisions should cover delays in installation and/or service, alarm response requirements, nonresponse conditions, and system testing requirements.

Your customer is an integral part of his system. He must test it periodically to determine that it is working properly. Your contract should specify when and how the system should be tested. It also should specify the procedure for requesting service if your customer suspects that a problem exists.

Limiting Liability

Alarm companies provide services for the public good. This is part of the reason that courts have upheld limits of liability clauses. If alarm companies charged fees based on the protected property's value and risk, only a few wealthy persons could afford alarm systems. So far, the courts have recognized that fees are based, in part, on the cost of equipment, operating expenses, and a margin of profit.

Other factors also play important roles in alarm companies' arguments to limit liability. Specifically, the entire system is not under your complete control. Although you installed a good system to provide complete protection, you cannot guarantee that your customer will use it properly. In fact, you cannot guarantee that he will use it at all.

You also cannot control the quality or reliability of telephone lines, a common alarm transmission component. If the telephone lines go down, you are not responsible and, therefore, should not be held liable if a loss occurs. The same argument holds for electric utility companies.

Nor can you guarantee police response. Your responsibility ends when you have notified the appropriate authorities that an emergency condition exists.

These are just a few of the possible occurrences that can affect your liability. Your contract should clearly define what services you are and are not responsible for.

Here's a final word to the wise: find a good attorney and work with him. You'll be glad you did.

Trends: Yesterday, Today, and Tomorrow 20

The alarm industry is old and it is new. Alarms have been used for more than 100 years, yet computer technology has given us the ability to offer sophisticated products and expanded services only within the last decade. Untold numbers of older systems, some installed thirty, forty, or fifty years ago, are still in operation. Hundreds or thousands of new systems are being installed daily.

Industry growth during the past ten years has been good. Even during the recessionary periods experienced during the late 1970s and early 1980s, most alarm companies recorded profits. During those slower economic times, many dealers may have had slowed income, but relatively few stalled or fell. Although no industry is recession proof, the alarm industry has shown that it is recession resistant.

One reason for this healthy economic record and forecast may be a heightened consumer awareness of the need for security. Consumers may not be consciously aware of it, but they are frequently exposed to the topic of security. News broadcasts often refer to security procedures used by the armed forces. Crime reporting is common in the media. Moreover, special crime prevention programs or articles are featured on television and radio as well as in newspapers and magazines.

Being successful in any business requires dedication. The alarm industry is no different. The Small Business Administration says that almost as many businesses fail each year as start operations. The alarm industry's failure rate is estimated to be 30%. That is, about three out of every ten alarm companies close their doors each year. As with most businesses, the causes of failure usually are undercapitalization and lack of management experience. A check of this year's *Yellow Pages* against last year's probably will reveal an increase in the number of alarm companies, however. More are starting than are failing.

CRIME PAYS

The more crime, the more the need for security—and the more business for aggressive alarm dealers. The shift of crime from urban to suburban and rural areas has opened new markets for alarm companies. Small businesses and homeowners are investing in security. Previously, most systems were installed for larger businesses and upper-income homeowners. As they have become more affordable for middle-class homeowners and small businesses, and the perception of need has increased, the overall market has shown almost limitless possibilities.

People are beginning to look not just for security, but for effective security. System up-grades are common and, if actively pursued, show promise as a profitable market segment. Crime may pay for criminals. It also can be profitable for those who are selling crime-prevention and protection products and services.

The most notable market change during the remainder of the 1980s and early 1990s will be alarm companies' customer mixes. Dealers who previously installed only commercial systems will begin to look at the expanding residential market. Those who installed only a few residential alarm systems will find their residential customers growing in numbers and as a percentage of their business.

Competitiveness

Increased competition is almost a certainty. Alarm dealers who used to be in no-bid situations will find themselves bidding against competitors, which may include telephone and cable companies in addition to other dealers. To remain competitive and profitable, they will have to review their marketing and pricing strategies more frequently. Quarterly reviews and revisions will not be uncommon.

Stories of price cutting and fierce competition already exist. Special deals, do-it-yourself systems, and whiz-bang alarm gadgets currently can be found in the market. Their existence will be short lived, however, and their overall impact on professional alarm dealers will be minimal.

Alarm dealers with good local reputations who offer reliable products and needed services will survive and will have profitable operations. The deals, packaged systems, and gadgets will fade because they do not provide true security. A $99.95 special cannot provide the features or the level of security of a professionally installed alarm system.

Room for the Little Guy. There will always be a place for the small alarm company. The small dealer can offer personal touches that larger companies, especially national organizations, cannot. Often, it is this personal touch and community identification that helps ensure successful competition against the larger companies.

With remote alarm monitoring services, such as those provided by Emergency 24 in Chicago and 3M in St. Paul, small companies can offer many, if not most, of the services provided by larger alarm companies that operate their own central stations. Already, the same high-tech products are available to small alarm dealers at affordable prices.

ADVANCED EQUIPMENT

Another trend that started in this decade, and will continue at least until the end of this century, is the influx of advanced, high-tech security products on the market. Security equipment is getting smarter, thanks to microprocessor technology.

You'll be seeing a plethora of new security devices in the coming years. Most of them will be designed to increase alarm systems' reliability and reduce false alarms. A few of the products will be based on technologies not previously used.

New radio frequency (RF) wireless systems will compete feature for feature with hard-wired systems and economies of scale will lower prices to dealers. Advances in RF wireless technology may possibly be the most significant development for the remainder of the century.

The 1980s have shown that passive infrared detectors are the most popular space protection devices for residential and commercial installations. Their popularity is not expected to wane in the foreseeable future.

Simply put, the future for alarm dealers looks good. You'll be able to offer your customers new products and services at prices that are competitive and profitable.

GOOD NEWS AND BAD NEWS

I'd like to wrap up this discussion of trends with a few words on the most significant issue facing the industry: false alarms. It is one of those "good news, bad news" stories.

Let's start with the bad news, which is something we all recognize: false alarms are a problem and they have been for more than a century. Now the alarm industry is being forced to face and solve the problem or suffer the consequences. The consequences are punitive false alarm ordinances and, more important, erosion of the industry's credibility.

The good news is that we, as a united industry, have the ability to reduce false alarms significantly. They will probably never be eliminated, but they can be reduced to acceptable levels. The key to this is in unity of purpose. Without a united effort, we'll only be marginally successful at best.

Alarm dealers and law enforcement agencies in Atlanta, Louisville, Cincinnati, and other cities have joined forces and have reduced the number of false calls to which police officers respond. If it can work in those communities, it can work in yours.

Your participation in the national campaign to reduce false alarms is encouraged and needed. If you are not part of the solution, you are part of the problem. You and your company will be affected and will suffer the consequences unless the problem is remedied quickly. Contact your local alarm association or the National Burglar and Fire Alarm Association today. You'll be glad you did. It's our industry. Therefore the solution is up to us.

SOURCES V

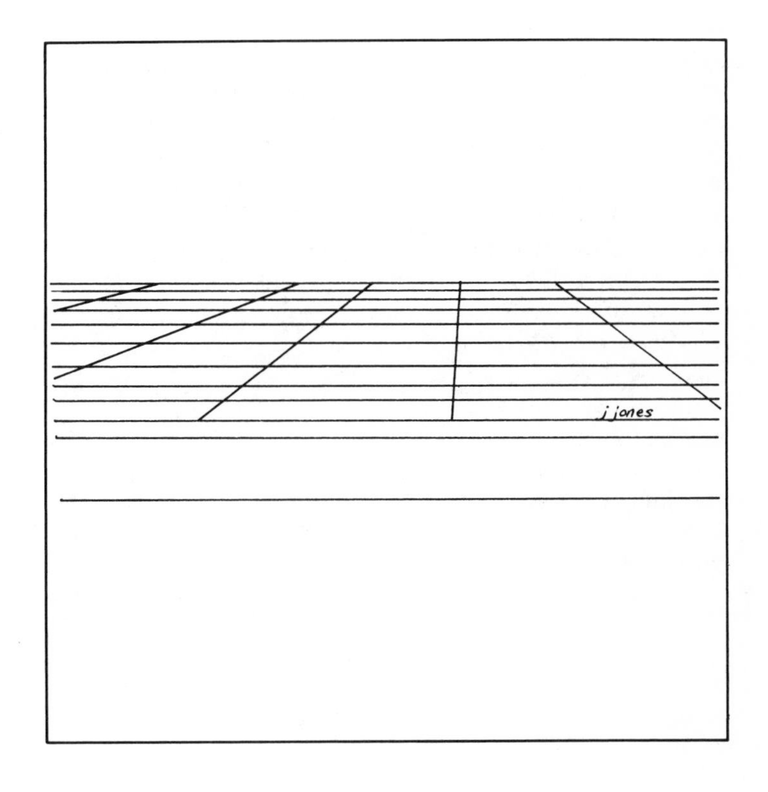

Appendix A Sources of Information

If you want to get ahead and stay ahead in the alarm industry, you must keep informed. An abundance of information exists if you know where to look for it.

Trade journals provide information to help you manage your business. They also offer insights into installation techniques and announce new products. Subscriptions usually are free to qualified subscribers, that is, those who are actively working as alarm dealers or installers. For subscription information, write to the journals at the addresses listed.

Numerous books are available. Many are invaluable reference tools that should be in every alarm dealer's professional library.

You also may need information about state licensing of alarm companies and state fire codes. A comprehensive list of state agencies is included to help you locate this important information.

TRADE JOURNALS AND BOOKS

Journals and Periodicals

Security Distributing & Marketing (SDM), Cahners Publishing Co., 1350 E. Touhy Ave., Des Plaines, IL 60018, (312) 635–8800

Alarm Installer & Dealer (AID), 5667 Las Virgenes Rd., Calabasas, CA 91302, (818) 880–5957

Security Dealer, 101 Crossways Park West, Woodbury, NY 11797, (516) 496–8000

Books

Butterworth Publishers, 80 Montvale Ave., Stoneham, MA 02180, (617) 438–8464

Alarm Dealer Licensing and Fire Code Information

State	Who to Contact for Alarm Dealers Licensing Information	Who to Contact for Fire Code Information
Alabama	Attorney General	State Fire Marshal (Department of Insurance)
Alaska	Legislative reference library	State Fire Marshal (Division of Fire Protection & Public Safety)
Arkansas	Board of Private Investigation & Private Security	

Alarm Dealer Licensing and Fire Code Information (*continued*)

State	Who to Contact for Alarm Dealers Licensing Information	Who to Contact for Fire Code Information
California	Bureau of Collection & Investigation	State Fire Marshal
Colorado	Department of Regulatory Agencies	Fire chiefs of individual districts
Connecticut	Department of State Police	State Fire Marshal (Commissioner of State Police)
Delaware	Attorney General	State Fire Marshal, State Fire Prevention Commission
District of Columbia	Business License Office	Fire Department
Florida	Department of State, Licensing Div.	State Fire Marshal (Treasury Insurance Commissioner)
Georgia	Attorney General	State Fire Marshal (Controller General)
Hawaii	Board of Private Detectives & Guards	State Fire Marshal (Director, State Department of Regulatory Agencies)
Idaho	Attorney General	Department of Labor
Illinois	Department of Registration & Education	State Fire Marshal (Department of Law Enforcement)
Indiana	Indiana State Police	State Fire Marshal
Iowa	Bureau of Criminal Investigation	State Fire Marshal (Department of Public Safety)
Kansas	Attorney General	State Fire Marshal
Kentucky	State Police	State Fire Marshal (Department of Insurance, Consumer Protection Regulation Cabinet)
Louisiana	Attorney General	State Fire Marshal
Maine	Department of State, Corporate Division	State Fire Marshal (Department of Public Safety)
Maryland	State Police	State Fire Marshal (Department of Public Safety & Correctional Services)
Massachusetts	Department of State Police, Records	State Fire Marshal (Department of Public Safety)
Michigan	Department of State Police, Records	State Fire Marshal (Director, Department of State Police)
Minnesota	Department of Public Safety	State Fire Marshal (State Commissioner of Public Safety)
Mississippi	Attorney General	State Fire Marshal (State Insurance Commissioner)
Missouri	Attorney General	State Fire Marshal (Department of Public Safety)
Montana	Attorney General	State Fire Marshal (Attorney General)
Nebraska	Attorney General	State Fire Marshal (Director, Department of Insurance)
Nevada	Attorney General	State Fire Marshal Division (Department of Commerce)

Alarm Dealer Licensing and Fire Code Information (*continued*)

State	Who to Contact for Alarm Dealers Licensing Information	Who to Contact for Fire Code Information
New Hampshire	Department of Safety	State Fire Marshal (Director of Safety Services)
New Jersey	Department of Law & Public Safety	Local Fire Departments
New Mexico	State Construction Industries Agencies	State Fire Marshal (Superintendent of Insurance)
New York	Department of State	State Labor Department
North Carolina	State Bureau of Investigations	State Fire Marshal (Commissioner of Insurance)
North Dakota	Attorney General	State Fire Marshal (Attorney General)
Ohio	Department of Commerce, Licensing Division	State Fire Marshal (Department of Commerce)
Oklahoma	Private Detective & Security Agencies	State Fire Marshal Agency
Oregon	Department of Commerce	State Fire Marshal (Department of Commerce)
Pennsylvania	Court of Quarter Sessions	State Fire Marshal (State Police Bureau of Fire Protection)
Rhode Island	Department of Business	State Fire Marshal (Division of Fire Safety, State Executive Department)
South Carolina	Secretary of State	State Fire Marshal (Department of Public Safety)
Tennessee	Department of Revenue	State Fire Marshal, Division of Fire Prevention (Commissioner of Insurance)
Texas	Board of Private Investigators	Texas Fire Marshal Department, State Fire Marshal (State Board of Insurance)
Utah	Secretary of State	State Fire Marshal (State Fire Prevention Board)
Vermont	Secretary of State	State Fire Marshal (State Police, Bureau of Criminal Investigation)
Virginia	Attorney General	State Fire Marshal (Division of State Corporation Commission)
Washington	Attorney General	State Fire Marshal (State Insurance Commissioner)
West Virginia	Secretary of State	State Fire Marshal (State Insurance Commissioner)
Wisconsin	Department of Regulation & Licensing	State Fire Marshal (Administrator, Division of Criminal Investigation)
Wyoming		State Fire Marshal (Department of Fire Prevention & Electrical Safety)

Appendix B Professional Associations

Growing professions form active associations to tackle their industries' problems and promote professionalism. The alarm industry has numerous local, regional, and national associations. Membership in one or more of these helps to keep you informed and to ensure your success in the industry.

The following pages provide a listing of professional associations. Find the ones for your area and meet with them.

National and Regional Alarm Associations

Region	Association	Contact
National	Alarm Association of America	Howard Owens, Chief Executive Officer Police Alert Security Systems P.O. Box 1122 Belleville, IL 62223 (618) 274-9445
National	Central Station Electrical Protection Association (CSEPA)	Robert J. Daugherty, Executive Secretary CSEPA 1120 Nineteenth St., N.W. Suite LL20 Washington, DC 20036 (202) 296-9595
National	National Burglar and Fire Alarm Association (NBFAA)	Robert J. Daugherty, Executive Secretary NBFAA 1120 Nineteenth St., N.W. Suite LL20 Washington, DC 20036 (202) 296-9595
National	Computer Alarm Association	Ed Morawski, President Honeywell/American 1131 Race Street Cincinnati, OH 45210 (513) 241-5511
National	National Fire Protection Association	Robert W. Grant, President Batterymarch Park Quincy, MA 02269 (617) 770-3000
New England	New England Alarm Association Council	Dave Armstrong, President 1 Mount Vernon Street Winchester, MA 01890 (617) 891-7900
Western	Western Burglar and Fire Alarm Association (WBFAA)	Les Gold 8500 Wilshire Boulevard Beverly Hills, CA 90211 (213) 652-0490

State Alarm Associations

State	Association	Contact
Alabama	Alabama Alarm Association	James Yeaman, Executive Director P.O. Box 4599 Montgomery, AL 36103 (205) 834-2001
Arizona	Arizona Burglar Alarm Association	Galen Buckey, President Sonitrol Security Systems 545 Osborn Phoenix, AZ 85012 (602) 233-1533
California	East Bay Alarm Association	Ken Rollegh, President Alarmex P.O. Box 6003 Hayward, CA 94540 (415) 785-0141
	Inland Empire Alarm Association	Precision Security Systems, Inc. 9980 Indiana Avenue, Suite 6 Riverside, CA 92503 (714) 780-2700
	Mid Cal Alarm Association	Central California Alarm Company 1271 N. Wishon Fresno, CA 93728 (209) 222-2741
	Burglar and Fire Alarm Association of Orange County	Dick Beunk, President Comseco 1227 W. Collins Orange, CA 92667 (714) 997-1381
	Coachella Valley Burglar and Fire Alarm Association	Steve Kaufer, President Checkmate Security Systems P.O. Box 1566 A Palm Springs, CA 92663 (609) 320-6941
	Sacramento Alarm Association	Harold Anderson, President Antru Security Sacramento, CA 94825 (916) 485-7447
	Burglar and Fire Alarm Association of San Diego County	Daniel Greenwald, President California Coast Alarms, Inc. 4425 Convoy San Diego, CA 90111 (619) 560-0333
	Los Angeles Burglar and Fire Alarm Association	Bill Joyce, President West Coast Alarm Systems 2326 Sawtelle Boulevard Los Angeles, CA 90064 (213) 231-9204
Colorado	Colorado Burglar and Fire Alarm Association	United Security Systems 260 S. Raritan Denver, CO 80223 (303) 778-7000

State Alarm Associations (*continued*)

State	Association	Contact
Connecticut	Connecticut	Pete Orvis, President P. W. Orvis & Associates 51 Littlebrook Road Wilton, CT 06897 (203) 762-8429
Delaware	Delaware Alarm Association	McCorkle Alarm Company Wilmington, DE 19805 (302) 656-4326
Florida	Alarm Association of Florida 1271 SW 28 Way Ft. Lauderdale, FL 33312 (305) 587-8781	Floyd F. Neely, Executive Director Systems for Security 3010 NW Seventeenth Avenue Miami, FL 33142 (305) 633-0613
Georgia	Georgia Burglar and Fire Alarm Association	John Thomas, Executive Director 1795 Peachtree Road, NE Suite 208 Atlanta, GA 30309 (404) 874-9168
Hawaii	Hawaii Burglar and Fire Alarm Association	Ed Alfonso, President Alarm Guard 2952 Koapaka Street Honolulu, HI 96819 (808) 833-5579
Illinois	Illinois Burglar and Fire Alarm Association	Cunningham Security Systems 675 W. Ardmore Roselle, IL 60172 (312) 579-0640
Indiana	Indiana Burglar and Fire Alarm Association	Able Alarms 200 Washington Street Kokomo, IN 46901 (317) 452-3232
Kentucky	Kentuckiana Burglar and Fire Alarm Association	Bluegrass Security Systems 132 Chenowith Lane Louisville, KY 40207 (502) 897-2303
Louisiana	Alarm Association of Greater New Orleans	Richard Carrigee, President Main Electronics 2318 David Drive Metairie, LA 70003 (504) 887-8540
Maine	Maine Burglar and Fire Alarm Association	Seacoast Security Systems P.O. Box 734 Camden, ME 04843 (207) 236-4848
Maryland	Mid Atlantic Alarm Security Association	Dallas Johnson, President Dallas Johnson Dictograph Security Systems 4850 Rugby Avenue Bethesda, MD 20814 (301) 652-1990

State Alarm Associations (*continued*)

State	Association	Contact
Massachusetts	Massachusetts Security Contractors Association	Susan Cavichi, Executive Director 1 Mount Vernon Street Winchester, MA 01890 (617) 729-2335
Michigan	Burglar and Fire Alarm Association of Michigan	Rolland Davis, President Midnight Burglar Alarm Systems, Inc. P.O. Box 3459 Flint, MI 48503 (313) 232-3171
Minnesota	Minnesota Burglar and Fire Alarm Association	Metro Alarm, Inc. 13724 Nicollet Burnsville, MN 55337 (612) 894-6960
Mississippi	Mississippi Alarm Association	Guardian Alarm Systems 205 Dogwood Drive Starksville, MS 39754 (601) 323-2430
Missouri	Alarm Association of Greater St. Louis	Al Cobb, President National Alarm Company 2001 South Big Bend St. Louis, MO 63117 (314) 647-0064
Nebraska	Nebraska Association of Alarm Companies	Security International 4622 S. 88th Omaha, NE 68127 (402) 331-6050
Nevada	North Nevada Alarm Association	Alarmex P.O. Box 4256 Reno, NV 89431 (702) 358-5221
New Hampshire	New Hampshire Alarm Association	Pope Security Systems Caroline Road Silver Lake, NH 03875 (603) 367-4397
New Jersey	New Jersey Burglar and Fire Alarm Association	Liberty Security Systems P.O. Box 105 Howell, NJ 07731 (201) 759-0303
New York	Hudson Mohawk Alarm Association	Jim Hart, President Hart Alarm Systems 12 Fairview Road Loudonville, NY 12211 (518) 462-2313
	Metropolitan Burglar Alarm Association	Brenda Mahler, Executive Director 333 N. Broadway, Suite 2000 Jerico, NY 11703 (516) 433-6006
	New York Burglar and Fire Alarm Association	Bernard Shore, Administrator Kerman Protection

State Alarm Associations (*continued*)

State	Association	Contact
New York (*cont.*)		42-02 192 Street Flushing, NY 11358 (212) 352-1700
North Carolina	North Carolina Alarm Association	Marcie Hege, Executive Director P.O. Box 1630 Raleigh, NC 27602 (919) 362-7109
Ohio	Akron Alarm Association	Guardian Alarm Services, Inc. 1349 East Avenue Akron, OH 44307
	Alarm Advisory Council of Central Ohio	A&T Security and Video Systems, Inc. 2850 Fisher Road Columbus, OH 43204 (614) 272-5318
	Greater Cincinnati Regional Chapter #1	Tony Cioffi, President Honeywell/American 1131 Race Street Cincinnati, OH 45210 (513) 241-3040
	Greater Dayton Burglar and Fire Alarm Association	Sonitrol of Southwestern Ohio 1430 East 3rd Street Dayton, OH 45403 (513) 228-7301
	Lake Erie Burglar and Fire Alarm Association	Fidelity Alarms, Inc. Wickliffe, OH 44092 (216) 585-9500
Oklahoma	Oklahoma Alarm Association	Johnnie Fletcher, President Protection Alarm Company 1506 Linwood Oklahoma City, OK 73106 (405) 272-0781
Oregon	Oregon Burglar and Fire Alarm Association	Preloc Security Systems, Inc. 4605 NE Fremont Portland, OR 97213 (503) 281-1415
Pennsylvania	Pennsylvania Burglar and Fire Alarm Association	Pat Egan, President Commonwealth Security Systems, Inc. 20 East Walnut Street P.O. Box 97 Lancaster, PA 17602 (717) 394-3781
Rhode Island	Alarm Association of Rhode Island	Joseph Trillo, President AAA Custom Alarm Systems 642 East Avenue Warwick, RI 02886 (401) 826-0800
Tennessee	Tennessee Burglar and Fire Alarm Association	Ed Kerley, President Professional Security Concepts

State Alarm Associations (*continued*)

State	Association	Contact
Tennessee (*cont.*)		P.O. Box 10794 Knoxville, TN 37919 (615) 524-8300
Texas	Dallas County Burglar and Fire Alarm Association	Chester Jones, President Texas Security Central, Inc. P.O. Box 129 Addison, TX 75001 (214) 351-0130
	Greater Houston Alarm Association	Southwestern Security 2710 Bissonnett Houston, TX 77005 (713) 523-6713
	Greater Fort Worth Alarm Association	Ed Schwartz, President Armco Burglar Alarm Co., Inc. 7160 Baker Boulevard Fort Worth, TX 76118 (817) 589-1234
	Texas Alarm and Signal Association	Larry Sondock, President McCane-Sondock Protection Systems, Inc. P.O. Box 52991 Houston, TX 77052 (713) 654-9252
	Texas State Association of Security Professionals	Jim Sinclair, President Saturn Security Systems, Inc. 9931 Harwin, Suite 140 Houston, TX 77036 (713) 781-0788
Vermont	Vermont Alarm and Signal Association	Marshall Lock and Alarm Middlebury, VT (802) 388-7633
Virginia	Virginia Burglar and Fire Alarm Association	Lawrence Bond, President Southeastern Security Systems P.O. Box 687 Portsmouth, VA 23705 (804) 393-6644
Washington	Burglar and Fire Alarm Association of Washington	Sound Security, Inc. P.O. Box 2551 Everett, WA 98203 (206) 258-3655
Wisconsin	Wisconsin Burglar and Fire Alarm Association	Glenn Lauren, President Bonded Alarm 4712 W. Burleigh Milwaukee, WI 53210 (414) 442-9999

Appendix C Glossary

Abort: a feature of some tape dialers and digital communicators that discontinues the transmission of an alarm signal if the alarm system is turned off. Helpful in reducing false alarms.

AC: alternating current.

Access control: any means of limiting entry to authorized personnel.

Account: a subscriber to an alarm company's services.

Active detector: a detector that functions by transmitting or sending out energy. Examples: microwave, ultrasonic, photoelectric beam.

Air turbulence: air disturbance caused by a breeze or draft from sources such as fans, air conditioners, or furnace vents. May cause false alarms in the vicinity of some detectors.

Alarm condition: an occurrence such as an intrusion, fire, or hold-up that is detected and signaled by an alarm system.

Alarm control: a device that provides power to operate an alarm system and allows the system to be turned on and off.

Alarm indicator: an audible or visible device used to indicate an alarm condition.

Alarm line: a wire or telephone line used to report an alarm condition to a remote location.

Alarm screen: a window screen with fine wires woven into the fabric and connected to the protective loop so that cutting or removing the screen will trip the alarm.

Alarm signal: a signal that a condition such as an intrusion, fire, or hold-up is occurring.

Alarm system: a combination of devices that detects and reacts to an abnormal condition.

Alignment: the physical aiming of photoelectric beams or motion detectors; or the electronic tuning of radio frequency circuits.

Alternating current: an electric current that continuously reverses direction.

Ampere (amp): a unit of measure; the rate of electrical flow.

Ampere-hour: the capacity of a battery.

Annunciator: a visible and/or audible indication, such as a siren, light, or bell, of an alarm or other special condition.

Area protection: *see* Volumetric protection.

Armed: the condition of an alarm system when it is turned on, ready to trip when a detector is activated.

Armed light: a light or LED that indicates the alarm system is armed.

Attack: an attempt to burglarize or vandalize; or an attempt to defeat an alarm system.

Audible alarm: also known as a local alarm; an alarm signal that can be heard, such as a siren or bell.

Audio discriminator: a device that detects noise. Some models trigger an alarm only on certain sound frequencies, on sounds above a certain level, on sounds lasting a certain length of time, or a combination of these.

Automatic reset: a feature of some alarm systems that automatically silences the annunciator and returns the system to its nonalarm condition after a certain length of time.

Automatic shutoff: a feature of some alarm systems that automatically silences the annunciator after a certain length of time.
BA: burglar alarm.
Battery: the power source generally used for standby power in alarm systems.
Bell: a noise-making device sometimes used as an annunciator in alarm systems.
Bug: a detection device; to install such a device.
Burglar alarm: *see* Intrusion alarm.
Burglary: the entry into a building, forced or otherwise, with the intent to commit a theft or any felony.
Burnishing tool: a very fine tool used to clean relay contacts.
Cable: a bundle of wires covered by a common jacket. There can be thousands of conductors in a cable.
Capacitance detector: a device that detects an intruder touching or closely approaching a protected metal object, usually a safe or file cabinet. Also called a safe alarm or proximity alarm.
Carrier current system: a type of alarm system that has a receiver and transmitters plugged into the 120-volt house wiring. The high-frequency alarm signal is transmitted through the house wiring.
Casement window: a window that opens outward, usually operated with a crank.
CCTV: closed-circuit television.
Central station: a central location where an alarm company monitors its accounts. Upon receiving an alarm signal, the central-station personnel take whatever action is called for, such as notifying police and/or fire departments. In some cases, a guard from the alarm company is sent to the customer's premises.
Certified (UL certified): to be certified by Underwriters Laboratories, an alarm system must meet UL specifications for equipment, installation, and service.
Circuit: an arrangement of electrical components connected by wire.
Class II: low-voltage, limited-energy electrical systems.
Class II transformer: a transformer that limits energy. Such transformers are almost always used in supplying power for alarm systems.
Closed-circuit loop: *see* Single closed loop.
Closed-circuit television: an on-premises television system used as a monitor. System consists of a television camera, video monitor, and a coaxial cable connecting the two.
Coaxial cable: a special type of insulated, shielded cable, mostly used in CCTV systems.
Coded transmitter: a type of transmitter, such as digital dialer, specially coded to identify the origin and/or type of an alarm.
Color code, resistor: the color code that identifies the value of a resistor:

Black	0	Blue	6
Brown	1	Violet	7
Red	2	Gray	8
Orange	3	White	9
Yellow	4	Silver	10% tolerance
Green	5	Gold	5% tolerance

The first two colors indicate the first two digits; the next color indicates the multiplier, and the fourth indicates tolerance. Tolerance is 20% if there is no fourth color band.
Commercial alarm: an alarm system installed in a business.

Common: the terminal that completes an electrical circuit with a normally open or normally closed contact; or the terminal where many wires are connected.
Compromise: to defeat an alarm system.
Conduit: a type of pipe used to contain wires. Sometimes used for alarm system wiring.
Contact: a type of switch, either magnetic or mechanical, used in alarm systems. Usually used to protect doors or other openings.
Continuity: a continuous, uninterrupted electrical circuit.
Control: *see* Alarm control.
Control key switch: a key-operated switch on an alarm control unit used to turn the system on and off.
Copper: most wires used in alarm systems are copper.
Cord trap: a mostly outdated means of detection consisting of a cord or wire stretched across a doorway and connected to the protective circuit. Breaking or pulling the cord loose breaks the circuit and trips the alarm. Space-protection devices and photoelectric beams have replaced such traps in most modern alarm systems.
Crossover foil: thin, tinned brass foil or Mylar tape used to connect foil tape over crossbars on windows.
Current: the flow of electric charge through a circuit, measured in amperes.
Day-night control: a control unit that can be wired with two protective circuits. The day circuit is on twenty-four hours a day, while the night circuit is only on when the control is turned on.
DC: direct current.
Dealer: usually refers to an alarm company, one who sells and/or leases alarm systems.
Decibel (dB): a measure of loudness.
Dedicated circuit: a telephone line used solely for the transmission of alarm signals connecting a protected premises to an alarm monitoring facility.
Defeat: to avoid being detected by an alarm system, or to render the system inoperable.
Detector: any device that senses the presence of a dangerous, undesirable, or abnormal condition.
Deterrent: anything that discourages. A potential intruder should be discouraged by the presence of an alarm system.
Dialer: *see* Digital communicator and Tape dialer.
Digital communicator: also known as a digital dialer. A device for transmitting alarm signals over telephone lines using a communicator that sends a coded message to a receiver, usually at a central station.
Digital receiver: the device that receives, identifies, and processes the signals sent by a digital communicator.
Diode: an electronic device that allows current to flow only in one direction. Also known as a rectifier.
Direct connect: an alarm system that is connected directly to a police and/or fire department.
Direct current: an electric current that flows only in one direction. Batteries supply direct current.
Direct wire: an alarm system with a single dedicated circuit to send an alarm signal from the protected premises directly to the central station, where the signal is processed by a separate receiver.
Disarm: to turn off an alarm system.
Distributor: a person or company who sells many types of alarm equipment to alarm dealers.

DMM (digital multimeter): a very sensitive testing instrument that displays readings with numbers.

Door cord: a sturdy, flexible cord with a terminal on either end used to carry an alarm circuit to and from contacts on doors and movable windows.

Doppler effect: a shift in the frequency with which waves from a given source reach an observer. Microwave and ultrasonic devices use this principle to detect movement.

Double-hung window: a type of window often used in older buildings. Both the upper and lower sections of the window are movable.

DPDT: double-pole–double-throw switch.

DPST: double-pole–single-throw switch.

Dry cell: a nonrechargeable type of battery.

Electric eye: the light-sensitive device used in photoelectric systems.

Electronic key switch: a switch operated by pushing buttons in a certain sequence or combination.

Electronic siren: a speaker with a built-in electronic siren driver.

Electronic warbler: a speaker with a built-in electronic siren driver that produces a much different sound than that of an electronic siren.

Emergency button: also known as a panic button. A pushbutton switch that triggers an alarm system and activates the annunciator, usually whether or not the alarm system is turned on at the time.

Entry/exit delay: this feature allows the user to turn the system on and exit the premises (usually through one specific door) within a certain time period without activating the alarm. A time delay also allows the user to enter and turn off the system without activating the alarm if he does so within a certain time period. Entry/exit delays are becoming almost standard features on newer alarm systems, especially residential systems.

EOL: end of line.

EOL resistor: a resistor placed across the end of a protective loop.

False alarm: an alarm signal that is not caused by an intruder or other dangerous or undesirable condition. False alarms may be due to telephone line or equipment failure, but frequently are caused by the user.

Fire alarm: an alarm system that detects and reacts to fire.

Foil connector: used to connect foil to alarm circuit wire. Also called foil block or take-off block.

Foil tape: a ribbon of thin lead foil, applied to glass and connected to the protective circuit. The foil tape will break if the glass is broken, thus breaking the circuit and tripping the alarm.

Foot rail: a foot-activated switch, usually used to signal a hold-up.

Form A contact: single-pole–single-throw, normally open relay or momentary switch contact.

Form B contact: single-pole–single-throw, normally closed relay or momentary switch contact.

Form C contact: single-pole–double-throw relay or switch contact.

Fuse: an electrical safety device designed to open the circuit when dangerously high current could damage other equipment.

Gauge: a measure of the size of electrical wire. The higher the gauge number, the thinner the wire.

Gel cell: trade name of a lead acid battery. Also used for any battery that has a gelled electrolyte rather than a liquid electrolyte.

Glass-breakage detector: a window-mounted device that detects the breaking of glass by responding to the high frequencies caused by the glass breaking.

Ground: a wire or conductor that is in direct electrical contact with the earth. Cold water pipes are good ground connections. Also, "ground" may refer to a common point in an alarm system.

Hang-up signal: a signal programmed into a tape dialer that causes the dialer to "hang up" (disconnect from the phone line) at the end of the voice message.

Hard-wire: an alarm system in which the components are connected with wire, as opposed to a system that uses wireless transmitters.

Hold-up alarm: a silent alarm system used to notify authorities that a hold-up is in progress. It is always activated by inconspicuous devices, such as a foot rail, money clip, or hidden button.

Horn: a noise-making device used as an annunciator.

House circuit: protective circuit.

Indicator light: a light that indicates the status of an alarm system.

Infrared (IR) detector: there are two types. A passive infrared (PIR) device does not emit IR energy; rather, it measures existing IR energy. It detects an intruder by sensing body heat. An active infrared device is a photoelectric beam that emits IR energy instead of visible light.

Insulation: a material that does not readily conduct electricity. The plastic coating on wire is an insulator that prevents the wires from touching and conducting electricity between them.

Interior: within the premises.

Intrusion alarm: a combination of devices that detects and reacts to the presence of an intruder or the attempt to break into a protected location.

IR: infrared.

Keyed alarm control: the alarm system is turned on and off with a key from the exterior of the premises.

Lacing: fine wire stretched back and forth across skylights, ducts, or other openings. An intruder breaking a wire will trip the alarm. Most modern alarm systems use motion detectors instead of lacing.

Lamp cord: two-conductor, parallel construction, usually 18-gauge wire, typically used for lamps and extension cords. Often used in alarm systems.

Latching relay: mechanical—a relay with two coils. Momentarily energizing one coil operates the contacts, which mechanically latch and remain in position when the coil is deenergized. Contacts are operated in the opposite direction when the second coil is momentarily energized. Magnetic—a relay that latches by means of a magnet. Can be released by energizing the second coil or by energizing the first coil with reverse polarity DC current.

Lead acid battery: a type of rechargeable battery used for standby power in alarm systems.

Leased alarm: usually, a type of service in which the customer does not buy the alarm system. An alarm company generally leases equipment and provides a protective service.

Leased line: a line or circuit that is leased from the telephone company and used for alarm reporting.

LED: light-emitting diode.

Light-emitting diode: a small electronic device that gives off light when current is passed through it. Most often used as indicators in alarm systems.

Line cut monitor: a device that detects and reacts to a telephone line being cut or shorted.

Line seizure: a feature of some alarm systems that causes a tape dialer or digital communicator to disconnect all telephones from the line when the system is tripped, thus preventing an intruder from defeating the system by lifting the telephone receiver.

Line supervision: a feature of some alarm systems that has a certain current present on the line to the central station. Changing the current by cutting or shorting the line signals an alarm.

Listen-in: a feature of some alarm systems that allows central station personnel to listen for sounds of intrusion by means of microphones at the protected premises.

Local alarm: a loud alarm system, one that is designed only to make noise at the protected premises.

Loop-status indicator: a light, LED, or meter that indicates the status of a protective loop.

mA: milliamp.

Magnetic contact: a magnetically operated switch used on doors and windows to detect opening.

Maintained contact: a switch that is not momentary, it stays in its last set position.

Manual fire alarm box: a transmitter with a lever that activates a fire alarm when the lever is manually pulled.

Manual reset: a type of alarm system that must be turned off and physically restored to a nonalarm condition. Someone must go to the protected premises and reset the system, as opposed to an automatic reset.

Mat switch: a pressure-sensitive switch hidden under carpets and padding that trips an alarm when an intruder steps on it.

Medical alert: a type of alarm system that allows an invalid to notify someone that medical assistance is needed, usually by just pushing a button.

Mercury tilt switch: a switch that contains a small amount of mercury and is sensitive to position. When the switch is tilted, the mercury inside moves around and makes or breaks the circuit, thereby tripping the alarm.

Meter: an electrical device sometimes used to indicate the amount of current in a protective loop.

Micro-: a prefix meaning one-millionth (1/1,000,000).

Microwave detector: a device that detects movement by using the Doppler shift principle. The detector transmits and receives microwave energy that is reflected back by objects in the room. The movement of an intruder changes this pattern and trips the alarm.

Milli-: a prefix meaning one-thousandth (1/1,000).

Momentary contact: a type of switch that returns to its normal position when released; the opposite of a maintained contact switch.

Money clip: a special type of switch used in hold-up alarms. It is placed in a cash drawer, with the bottom bill of a stack inserted in the switch. The alarm is activated by removing that bill.

Motion detector: a device that detects movement within a protected area. Examples are microwave and ultrasonic detectors.

Multimeter: a testing device used to measure volts, ohms, and electrical current.

Multizone: an alarm system with two or more protective circuits connected to the same control unit.

National Electrical Code: the code specifying the safe use of electricity and safe wiring methods.

NC: normally closed.

NEC: National Electrical Code.

Ni-cad: nickel-cadmium battery.
Nickel-cadmium battery: a type of rechargeable battery used to supply standby power in alarm systems.
NO: normally open.
Normally closed: a switch or relay whose contacts are closed when nonenergized.
Normally open: a switch or relay whose contacts are open when nonenergized.
Normally open loop: a protective loop in which all detection devices are wired in parallel and are open in the secure mode.
Object protection: the protection of a single object, such as a safe or file cabinet.
Ohm: the unit of electrical resistance.
Ohm's law: the relationship between current and voltage: Amps = volts ÷ ohms.
Open: a break in a circuit.
Open circuit loop: *see* Normally open loop.
Opening: any possible point of entry—doors, windows, vents, skylights, etc.
Openings and closings: a prearranged schedule between an alarm user and the central station for turning the alarm system on and off at specified times.
Ordinance: a local law. Many cities are enacting new ordinances to regulate the use of alarm systems.
Overhead door: a truck, loading dock, or garage door. It usually requires specially made, wide-gap magnetic contacts.
Panic alarm: a loud alarm that is activated manually, usually with a pushbutton. Enables the user to trip the alarm whether the control is on or off.
Panic circuit: a twenty-four-hour circuit. When the panic device is activated, the alarm is tripped whether the system is on or off.
Parallel connection (batteries): connecting two or more batteries of the same type with the positive terminals connected together and the negative terminals connected together. This does not change the voltage, but reduces the current drain from each battery.
Parallel-pair wire: two wires that run next to each other and are connected with plastic insulation. *See* Zip cord.
Passive detector: a detector that does not emit or send out any type of energy in order to perform its detection function.
Passive infrared: *see* Infrared detector.
Pattern: the shape of the area or volume of coverage of motion detectors.
Perimeter protection: protection of openings on the exterior of a building, including doors, windows, vents, skylights, etc.
Pet mat: a special type of mat switch that is less sensitive to pressure, so that small pets will not trip the alarm by walking on the mat.
Photoelectric beam: a detection device that is activated by an intruder interrupting a beam of light by walking through it. Most modern units use infrared beams rather than visible light because infrared light cannot be seen.
Plunger switch: a type of switch that is operated by a plunger being depressed or released.
Polarity: the relationship of positive (+) and negative (−) terminals in DC circuits. It is important to be sure that alarm wires are connected to the correct terminals. If polarity is reversed, it could damage the system.
Power: the rate at which electrical energy is converted to another type of energy. Measured in watts.
Power supply: any source of electrical energy. Usually, an electronic device that converts AC to DC for use with alarm equipment.

Premises: the building or area protected by an alarm system.

Preventive maintenance: testing an alarm system on a regular basis to be sure that all components are functioning properly, preferably before false alarms or other problems occur.

Proprietary system: an alarm system that is owned and operated by the user, who is fully responsible for its operation.

Protected area: the portion of premises that is covered by an alarm system.

Protective circuit: also called protective loop. The electric circuit connecting all detection devices to the control unit.

Proximity alarm: *see* Capacitance detector.

Pushbutton: a momentary switch that is activated manually, usually used for panic buttons.

Radar alarm: an incorrect term sometimes used in reference to microwave detectors.

Radio frequency (RF): a transmission technology using radio waves for signaling status, supervisory and/or alarm information from one point to another.

Rate-of-rise detector: a type of thermal detector that responds to the rate of temperature change rather than to a fixed temperature.

Ready light: a light or LED used to indicate whether the protective circuit is in the secured condition.

Rechargeable battery: a battery that can be recharged, such as a lead acid or nickel-cadmium battery, as opposed to a dry cell battery that cannot be recharged.

Reed switch: a type of magnetic contact switch that contains small steel reeds in a glass tube.

Reflection: the property of materials to bounce back, or reflect, sound.

Relay: an electrically operated switch.

Remote arming: allows an alarm system to be turned on and off from any of several locations throughout the premises, away from the actual control unit.

Remote monitoring station: a facility such as a central station, police or fire department, etc., located away from the protected premises that receives and monitors signals.

Remote-zone annunciator: a device, usually a light or LED, located away from the control unit that indicates which portion of the system is in an alarm condition or unsecured.

Reset: to silence a loud alarm and/or to restore the system to a nonalarm condition.

Residential alarm: an alarm system installed in a private home.

Resistance: the opposition to current flow in an electrical circuit.

Ring-off: the sounding of the siren or bell in an alarm system.

Ripple: the result of imperfect filtering of a DC power supply operating from an AC source; the amount of AC power coming out of a DC power supply.

Safe alarm: *see* Capacitance detector.

Screen: *see* Alarm screen.

Seismic detector: an outdoor perimeter detector, usually buried in the ground, that senses the seismic waves of an intruder or vehicle moving over or near it.

Self-restoring alarm: a system that restores itself to a nonalarm condition after it has been tripped, if all detection devices have been returned to the secure position.

Self-stick foil: foil tape with an adhesive on one side.

Sensitive relay: an extremely sensitive type of relay that requires very little current to operate.

Series circuit: an electrical circuit in which all components carry the same current.

Service: the repair and/or maintenance of an alarm system.

Sensing device: *see* Detector.

Shelf life: the length of time an unused battery will remain in good condition.

Shielded wire: any wire or cable that is shielded with a metal covering.

Short circuit: a path for current flow in a circuit, usually caused by electrical contact between connecting wires.

Shunt: bypass.

Shunt switch: a switch used to bypass a detection device, usually at an entrance, to permit the user to enter without tripping the alarm system.

Signaling device: the annunciator of an alarm system.

Silent alarm: an alarm that does not sound at the protected premises, but transmits a signal to the central station or monitoring facility.

Single closed loop: a protective loop in which all detection devices are wired in series and closed in the secure mode.

Siren: a combination of speaker and siren driver, used as an annunciator in alarm systems.

Siren driver: an electronic device that generates the siren signal; a high-power speaker converts the signal to sound.

Slow-blow fuse: a type of fuse that is designed not to open as rapidly as a regular fuse.

Snaking: the procedure of running wire through walls using a flexible metal device called a snake.

Space alarm: *see* Volumetric protection.

SPDT: single-pole–double-throw switch.

Splice: a connection between two or more wires.

Spot protection: protecting one specific object or small area.

SPST: single-pole–single-throw switch.

Standby battery: a battery used to supply electricity to an alarm system in case of AC power failure.

Stranded wire: wire composed of many thinner wires twisted together.

Stripping: removing the insulation from the ends of wire to make connections.

Strobe light: a light with very rapid, bright flashes. Sometimes used as an annunciator in alarm systems.

Subscriber: a customer who leases an alarm system.

Supervised system: also called supervised loop. An open or closed circuit system that initiates an alarm if violated or indicates trouble with a circuit.

Swinger: an intermittent short or open condition in a protective circuit.

Switch: a device used to open and/or close electrical circuits.

Take-off block: foil connector.

Tamper switch: a switch used to detect the opening or moving of an alarm control or equipment box.

Tape dialer: a device for the transmission of alarm signals over telephone lines using a prerecorded tape message to notify someone at a remote location.

Telco: telephone company.

Telephone lines (dedicated): the method of transmitting signals from a protected premises to a remote monitoring station. These lines are leased from the local telephone company and are installed and maintained by the telephone company, but have no association at all with the user's existing telephone service.

Terminal: a wire connection point.

Test block: a terminal strip used for testing and troubleshooting.

Thermal detector: a device that senses a rapid rise in temperature or responds at a fixed temperature.

Time delay: a time interval, measured by an electronic circuit, used to provide a desired feature such as entry/exit delays.

Transformer: a device used to change the level of AC voltage.

Transmitter: a device that sends a signal to a remote point.

Trip: to activate or set off an alarm system.

Troubleshooting: the process of tracking down shorts, breaks, grounds, or other problems in an alarm system.

Trouble signal: a signal indicating a problem in the alarm system.

Twisted-pair wire: two wires that are held together by twisting around each other.

UL: Underwriters Laboratories, an independent testing agency.

UL standard for alarms: there are several UL standards applying to installation, equipment, and service that an alarm company is required to meet before obtaining a UL listing.

Ultrasonic detector: a device that detects movement using the Doppler shift principle; the detector transmits and receives high-frequency sound waves and reacts to changes in those sound waves.

Underdome bell: a type of alarm bell on which the hammer is completely covered by the gong.

Varnish: used on window foil after installation to protect the foil and help adhere it to the glass.

Vibration contact: a type of switch that responds to vibration of the surface it is mounted on.

Volt: a unit of electrical pressure; the force that drives current through a circuit.

Voltage drop: loss of voltage in circuit wires. The greater the distance voltage has to travel, the greater the voltage drop.

Volumetric protection: protection of three-dimensional space. Examples of such detectors are microwave, passive infrared, and ultrasonic.

V-O-M meter: volt-ohm-milliammeter. The basic alarm-testing tool.

Walk test: the testing of a motion detector by walking through the protected area to determine the range of the detector.

Watt: the rate of flow of electrical energy; the rate at which electrical energy is converted to other kinds of energy.

Window foil: foil tape.

Wireless: a method of sending alarm signals from small radio transmitters to a receiver, eliminating the need to run wires from detection devices to the control unit.

Zip cord: any parallel construction wire.

Zone: an individual area of a protected premises or individual detector in an alarm system, each zone having a separate indicator.

Zone light: a light or LED used to indicate the status of each zone in a zoned alarm system.

Zone shunting: removing a particular zone from an alarm system, allowing the rest of the system to function normally.

Appendix D Organization Manual

Not all parts of this manual will apply to your particular operation, but most are very basic and should be included in your manual. Whether your company is a sole proprietorship, partnership, or corporation, the various sections can be easily adapted to your operation.

It is suggested that each department or section manager have a copy of the organization manual so that all employees will have easy access to the information contained in it. Moreover, it is suggested that the manual be maintained in a three-ring, loose-leaf binder so that pages can be inserted and/or removed as the manual is revised. For the manual to be truly functional as a management tool, it should be reviewed periodically and revised as often as necessary to assure that its contents are current and accurate.

This sample should serve as a guide in preparing your manual. Some of the wording may need to be changed in some sections, depending on local, state, and federal legislation that affects your company's operations. (Explanatory notes by the author are shown in *italics.*)

Organization Manual for Magna Alarm Company*

The first page of the manual should be the title page showing the company name and Organization Manual.

Following the title page should be a general table of contents. Because the manual is in loose-leaf form so pages can be inserted and removed, page numbers should be omitted; however, the table of contents should indicate the sequence in which the various topics are presented.

Introduction

The purpose of this manual is to provide all employees with information that will give them an understanding and working knowledge of the functions, procedures, and policies pertaining to positions, departments, and divisions of the Company. The written policies, procedures, rules, and regulations contained in it are subject to continual revision. The manual will never be complete. Other policies, procedures, rules, and regulations will be issued periodically. Some will be new, while others will be amended versions of pages already in the book.

The philosophy of the Magna Alarm Company provides for the active participation and involvement of employees in the policy-making process. Recognizing that implementation of this philosophy will result in the presentation of the best possible ideas, the Company encourages participation in the area of policy development.

The introductory statement should include the purpose of the manual and encourage active employee participation.

Company Objectives

The objectives of Magna Alarm Company are:

1. To organize, staff, direct, coordinate, and control the sale and installation of alarm systems, and to provide quality service to our customers
2. To strive continually for quality in all company activities

Specific goals and objectives, such as annual sales goals, are not included in the Organization Manual; rather they are developed in management plans. The objectives stated in the manual are to provide general information to employees.

---

* Note: Magna Alarm Company and Four Forty-Four Systems, Inc., are fictitious companies.

Organization

Magna Alarm Company is organized as an operating division of Four Forty-Four Systems, Incorporated. As such, the Company is under the direction of the Vice President of Security Systems, Four Forty-Four Systems, Incorporated.

All staff functions will be performed at Magna Alarm Company headquarters.

The organization is structured so that the General Manager has sole authority for and is charged with the responsibility for the achievement of the Company's objectives.

The organization will be administered in such a manner that the highest possible productive efforts will be achieved by:

1. Expecting above-average performance from all personnel
2. Coupling responsibility with authority and reaching an understanding with those concerned before changing the scope of any responsibility
3. Providing that no person shall be given instructions by more than one person
4. Assuring that no difference of opinion between supervisors or employees as to authority or responsibility will be considered too trivial for prompt and painstaking attention
5. Assuring that supervisory personnel will counsel their employees and keep those employees informed of their progress and nature of their performance

This section establishes the framework for responsibility and authority that reaches throughout the organization.

Affirmative Action Policy

It is the Company's policy to employ the most qualified applicants. Accordingly, the Company will provide equal employment opportunity by recruiting, employing, assigning, compensating, evaluating, training, promoting, transferring, demoting, and terminating employees based on their own abilities, achievements, and experience without regard to race, color, religious preference, national origin or ancestry, marital status, sex, age, or economic status.

In support of this concept of employment on an equal basis, the Company will establish a plan of affirmative action to ensure compliance with this philosophy in practice as well as in spirit for all qualified persons.

Whether or not your company is required to meet the mandates established by the Equal Employment Opportunity Commission, it is a good idea to have an affirmative action policy. In addition to a statement similar to the one above, you should develop a specific, separate plan that outlines your efforts.

Recruitment Policy

To maintain the highest quality of employees to perform various tasks and jobs, the Company will obtain the most competent individual for each job.

Present employees are encouraged to apply for positions within the Company when they become available, if the employees qualify for those positions. Present employees will be considered on an equal basis with any outside applicants.

All available sources of manpower will be used. These will include but are not limited to: classified advertising in local newspapers, high school and college placement offices, local Veterans Administration office, and public and private employment agencies.

Organization Chart

An organization chart, graphically depicting authority and responsibility, should be included in the manual. Figures D.1 through D.4 show segments of an alarm company's organization chart. Whatever size of company, from a few employees to hundreds, such a chart is helpful in defining working relationships.

Job Descriptions

A job description for each position in your company should be included in the Organization Manual. The job description included here illustrates one format.

Installer

- Install alarm system components and equipment having critical operating requirements where installation, alignment, and adjustment of components, units, and completed system assembly/installation require a high degree of skill.
- Inspect components and inspect and test completed system.
- Do skilled installation tasks that require the use of secondary technical equipment for precise adjustment and alignment.

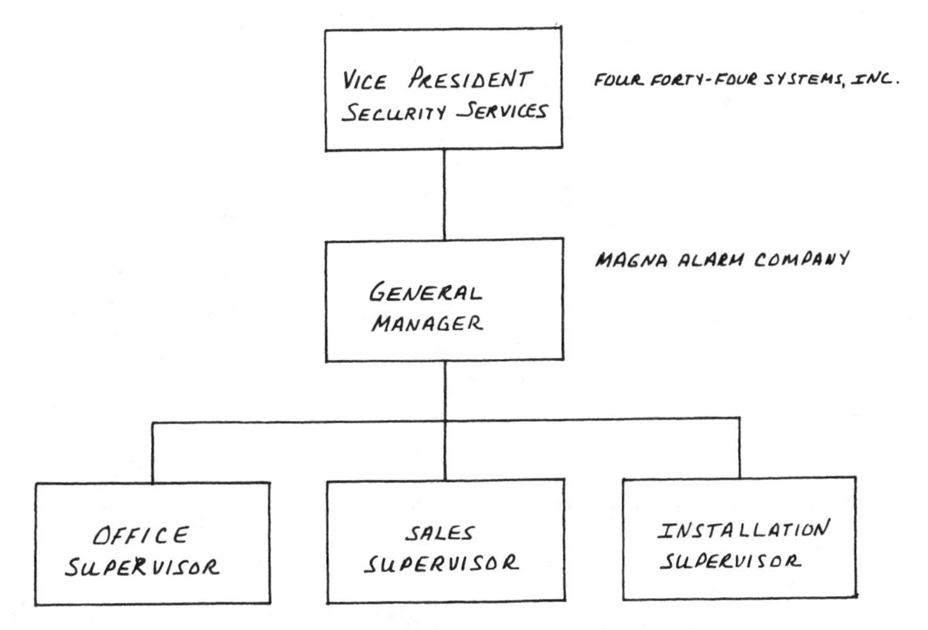

Figure D.1 Management staff organization chart.

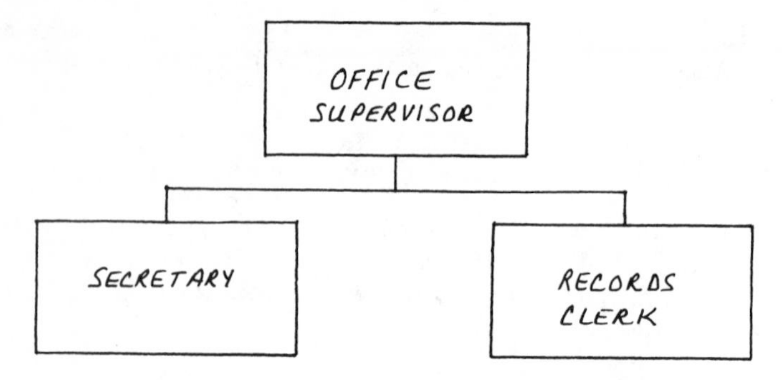

Figure D.2 Office staff organization chart.

- Work from complete and detailed plans and drawings.
- Select and use appropriate measuring instruments, calibration, testing, and inspection devices.
- Be able to proceed with minimum instruction and supervision.

Under the direction of and reports to the Installation Supervisor.
Education and experience requirements:

1. High school or vocational school graduate
2. Three years of general installation work including two years of alarm installation experience
3. One year of increasingly responsible supervision (general or alarm installation experience acceptable)

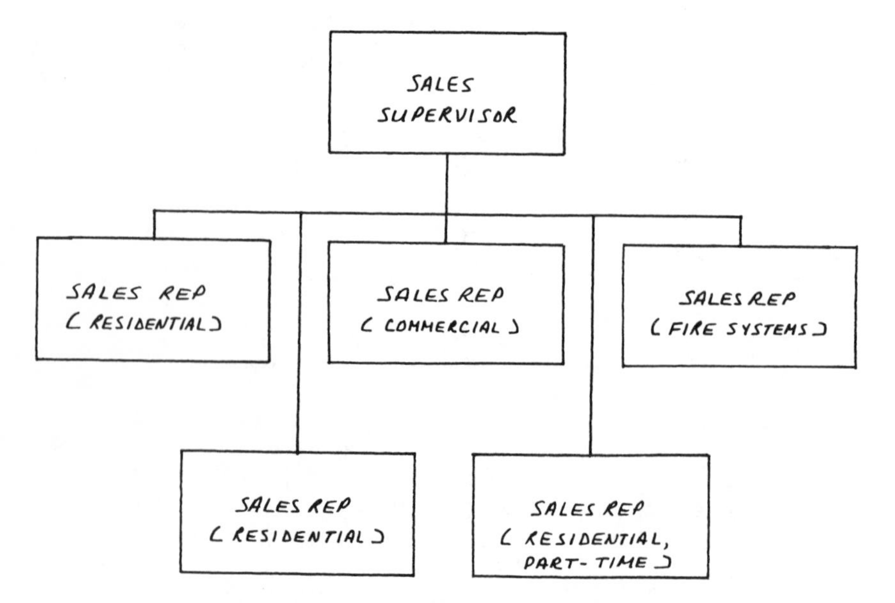

Figure D.3 Sales staff organization chart.

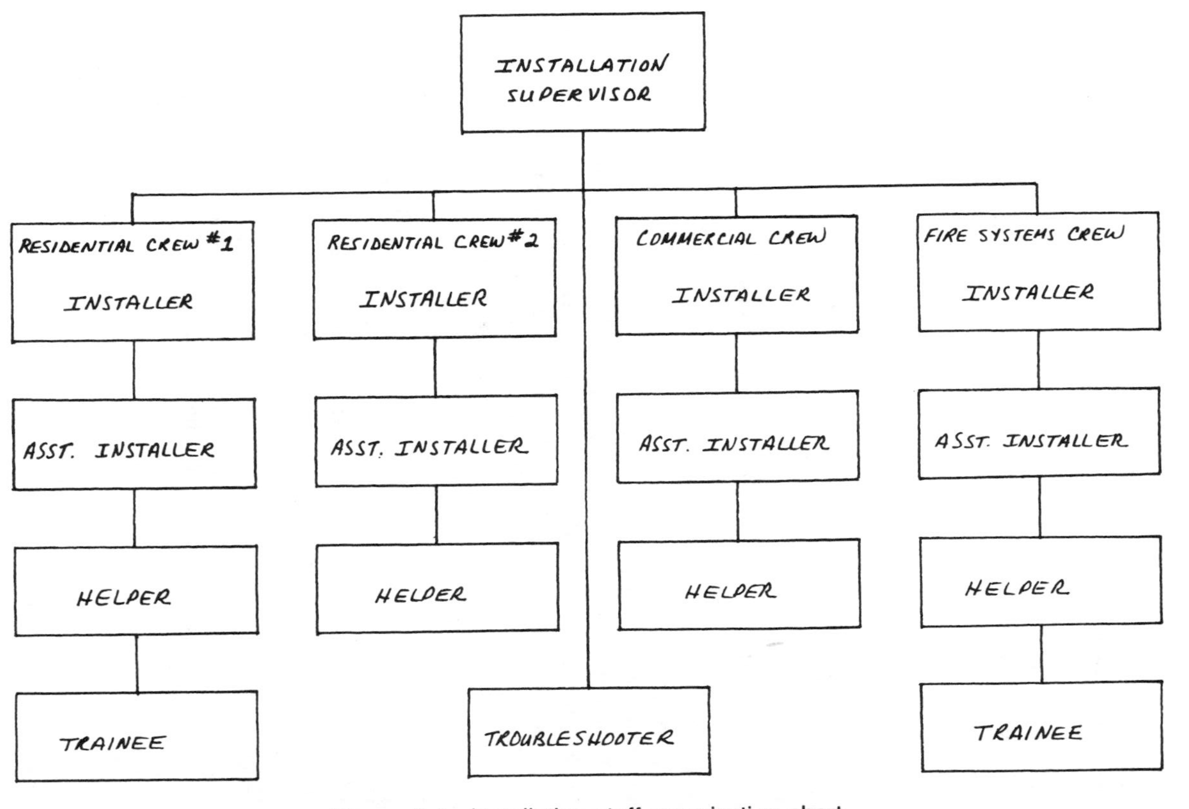

Figure D.4 Installation staff organization chart.

General Wage and Salary Policy

The Magna Alarm Company will pay wages and salaries that approximate the average of companies in its industry in its local geographic area.

Methods of Payment

Management and supervisory personnel will be paid on a monthly basis. The hourly rates given in the wage and salary schedule will be converted to a monthly wage. All management and supervisory personnel are considered exempt from overtime pay. All other personnel will be paid on an hourly basis and are entitled to overtime compensation.

Wage and Salary Schedule and Probationary Period

The salary schedule and accompanying policies and conditions will be applied to all employees without exception.

1. Entry into the salary schedule will be on step A except in those cases approved by the General Manager.
2. Every new employee will serve a probationary period of six working months. The probationary period will start on the first day of the month after employment (unless employed on the first day of the month). If employment of the individual is continued beyond the probationary period, he will be advanced to the next step on the salary schedule upon satisfactory completion of such period.
3. Each employee will have established permanent status and an anniversary date on the first day of the month after satisfactory completion of his probationary period.
4. One year after the employee's anniversary date, and each year thereafter, the employee will be advanced to the next step on the salary schedule.
5. After advancement to step E, the employee's performance will be reviewed and the employee will be placed in a pay range and step commensurate with his proved performance.
6. Merit raises, advancing to a higher range and/or step, may be granted by the General Manager based on recommendations by the employee's supervisor.
7. When an employee is promoted to a higher position, he will be placed on the first step of the new range on the salary schedule, which will grant a raise in salary of at least one step.

Authorized Paydays

Paychecks will be issued to all employees on the fifteenth and last days of each month. Should the fifteenth or last days of the month fall on a Saturday, Sunday, or holiday, the payday will be the last regular work day immediately preceding the fifteenth or last day of the month.

Salary Range Classifications

Position	Salary Range
Office supervisor	12
Secretary	6
Records clerk	2
Sales supervisor	10 + commission/override
Sales representative	6 + commission
Installation supervisor	22
Installer	14
Assistant installer	12
Installer's helper	9
Installer trainee	4
Troubleshooter	19

Salary Schedule

	Hourly Wage Rate				
Range	Step A	Step B	Step C	Step D	Step E
1	$ 4.75	$ 4.89	$ 5.13	$ 5.39	$ 5.66
2	4.99	5.14	5.40	5.67	5.95
3	5.24	5.40	5.67	5.95	6.25
4	5.50	5.67	5.95	6.25	6.56
5	5.78	5.95	6.25	6.56	6.89
6	6.07	6.25	6.56	6.89	7.23
7	6.37	6.56	6.89	7.23	7.59
8	6.69	6.89	7.23	7.59	7.97
9	7.02	7.23	7.59	7.97	8.37
10	7.37	7.59	7.97	8.37	8.79
11	7.74	7.97	8.37	8.79	9.23
12	8.13	8.37	8.79	9.23	9.69
13	8.54	8.80	9.24	9.70	10.19
14	8.97	9.24	9.70	10.19	10.70
15	9.42	9.70	10.19	10.70	11.24
16	9.89	10.19	10.70	11.24	11.80
17	10.38	10.69	11.22	11.78	12.37
18	10.90	11.23	11.79	12.38	13.00
19	11.45	11.79	12.38	13.00	13.65
20	12.02	12.38	13.00	13.65	14.33
21	12.62	13.00	13.65	14.33	15.05
22	13.25	13.65	14.33	15.05	15.80
23	13.91	14.33	15.05	15.80	16.59
24	14.61	15.05	15.80	16.59	17.42

The placement of positions in specific ranges will depend on your particular operation; however, the salary schedule should serve as a guide. Your salary schedule should be reviewed and revised periodically to keep pace with inflation.

Hours of Operation

The standard work day will be from 7:30 A.M. to 4:00 P.M. and the standard work week will be from Monday through Friday unless otherwise directed by the General Manager.

Overtime

Overtime will be allowed when specifically authorized in advance by the Department Supervisor or his designated representative in accordance with the rules and regulations of the Company.

Overtime will be authorized only for special jobs to which the employee is assigned. An eight-hour day and forty-hour week are recognized as the standards for all employees. Work authorized beyond these standards shall be compensated for at one and one-half times the regular rate of pay.

Employees who are required to perform special duties outside of the standard work day or work week will be guaranteed a minimum of two hours pay at one and one-half times their regular rate, providing such duties involve time that is not an extension of the regular work day and results either from the employee being summoned from home in an emergency or from a prearranged agreement.

The standard work month will be considered to have twenty-two days if a daily rate is established. Approved overtime earned in any month may be paid in the following month.

Rest and Meal Periods

All Company employees will be governed by the following:

1. Each employee will be entitled to a paid fifteen-minute rest break for each full four hours worked.
2. Employees working less than four hours will not be entitled to a paid rest break; however, a rest break may be granted without pay.
3. Employees working less than eight hours but more than four hours per day will be entitled to one fifteen-minute paid rest break. A second rest break may be granted without pay.
4. Employees working more than four hours will receive a meal period of thirty minutes. Those working less than four hours may be granted a meal period. All meal periods will be without pay.

Exclusion from Overtime

The law permits exclusion from overtime privileges for those employees whose positions are supervisory and whose duties, hours, and authority are such that time outside of the standard work day and work week is frequently involved.

Though excluded from overtime, the law further provides that such persons who, in the course of their employment, are required to work on a paid holiday will be paid at a rate equal to their normal rate of pay in addition to their regular pay for that holiday.

Paid Holidays

Most Company employees will receive time off with pay for the following holidays. Depending on their position with the company, some employees may be required to be on standby or on call for certain holidays, in which case those so designated will be entitled to double pay. The following holidays will be observed:

1. New Year's Day
2. Presidents' Day
3. Memorial Day
4. Independence Day
5. Labor Day
6. Thanksgiving Day
7. Day after Thanksgiving
8. Christmas Eve
9. Christmas Day
10. New Year's Eve

Should any of these holidays fall on a weekend, additional days will be designated. Employees observing recognized religious holidays other than those noted above will be granted time off with pay after permission has been granted by the General Manager. Whenever any of the previously noted holidays are federal holidays and observed on a day other than the actual holiday's date, the federal holiday schedule will be observed.

Vacations

Full-time employees who have completed their probationary periods are granted annual vacations with pay on the basis of length of service.

Length of Service	Vacation
Less than 1 year	½ day per month (maximum 5 working days)
More than 1 but less than 5 years	10 working days
More than 5 but less than 10 years	15 working days
More than 10 but less than 15 years	20 working days
More than 15 years	30 working days

All vacations must be approved by the employee's immediate supervisor and the General Manager.

Insurance Benefits

The Company will provide all employees with major medical coverage and life insurance. Included here are the highlights of the program; details are available from your supervisor. All premiums for this program will be borne by the Company.

Major benefits under the Major Medical package include:

$100 deductible
80% coverage after deductible is met
$20,000 maximum benefit per year
$50,000 maximum lifetime benefit

Life insurance benefits are as follows:

Personnel Covered	Coverage	Accidental Death & Dismemberment Coverage
Supervisors	$30,000	$30,000
All others	$20,000	$20,000

Additional coverages may be obtained at the employee's expense.
Premiums for dependents will be paid by the employee.

Payment of Insurance Premiums

Any employee and/or dependent insurance premiums paid by the Company will continue to be paid so long as the employee is eligible for salary, wages, sick leave, or vacation times pay benefits and consequently is actively on the Company payroll; or until such time as the individual's employment in the Company has been otherwise terminated. During any authorized leave of absence without pay, the employee may continue his insurance coverage at his own expense, providing such option is possible under terms of the appropriate insurance contracts.

Absence Due to Personal Illness

Each employee earns one day of sick leave for each full month of employment in the Company. Any employee who finds it necessary to be absent from work due to personal illness will notify his supervisor at the earliest possible opportunity. Such notice should be given not later than the beginning of the employee's shift on the day of absence.

When able to return, the employee will notify his supervisor before 4:00 P.M. on the work day preceding his return.

Accumulation of Sick Leave Time

Sick leave may be accumulated to a maximum of 100 days during the employee's uninterrupted employment. Approved leaves of absence shall not be deemed an interruption of employment.

Absences Other than Personal Illness

At the option of the employee, he may elect to take three of his personal sick leave days per year for the following reasons:

1. Death of a member of his immediate family.
2. Accident or emergency illness involving his person or property, or person or property of a member of his immediate family of such emergency nature that the immediate presence of the employee is required during his work day.
3. Appearance in court as a litigant or as a witness under an official order. (This section does not apply to those employees who have occasion to respond to alarm calls and, in the course of performing their duties, may become witnesses.)
4. Critical illness and/or surgery in the immediate family.
5. Necessary business leave (prior approval is required).

Definitions:

1. Immediate family is defined to include the mother, father, grandmother, grandfather, or a grandchild of the employee or the spouse of the employee, and the spouse, son, son-in-law, daughter, daughter-in-law, brother, or sister of the employee; or any relative living in the immediate household of the employee.
2. Emergency illness is defined as illness that occurs suddenly and without warning and requires the services of a physician.
3. Critical illness is defined as illness that is of such a serious nature as to require hospitalization, and the presence of the employee is deemed necessary for the patient's well-being and/or legal consultation.
4. Surgery is defined as a medical operation that requires the use of anesthesia, and the presence of the employee is deemed necessary for the patient's well-being and/or legal consultation.
5. Necessary business leave is defined as leave that is not social or recreational and cannot be conducted after the work day. Prior approval of the Department Supervisor is required.

The General Manager may require satisfactory proof for any such absence if it is deemed necessary.

Unauthorized absences of partial days for employees will be deducted on an hourly basis.

Additionally, an absence may be authorized, with sick pay benefits, for the following reasons (prior approval is required):

1. Religious holidays not recognized as holidays in the previous section on paid holidays
2. Bereavements other than immediate family
3. One-time special occasions for members of the immediate family:
   a. Graduation
   b. Special honors
   c. Military
   d. Marriage
4. Births, immediate family
5. Medical, immediate family
6. Education for employee—registration, consultation, and examination
7. Acts of God

Industrial Accident Leave

The Company carries state compensation insurance on all employees. If injured while working, the employee must report the injury, or see that it is reported, to his immediate supervisor on the same day that the injury occurs.

It is the responsibility of the supervisor who receives the report of injury to notify the office of the General Manager. This notice must be made within twenty-four hours of the occurrence of the injury so that the employee's insurance benefits will not be impaired.

In the event that the injury results in the employee receiving worker's compensation, the employee shall endorse to the Company any wage loss benefit checks received under the laws of the state. The Company, in turn, will issue to the employee his regular paycheck for the payment of wages or salary and will make the customary payroll deductions.

Retirement Plan

The Company's retirement plan provides qualifying participants with monthly lifetime income after retirement. All contributions to the retirement plan are made by the Company at no expense to the employee. Additional information is available from the General Manager.

Training

All training will be done on the job. It is the responsibility of supervisors to see that all personnel are properly trained. The Office Supervisor is charged with the responsibility of personnel function and will assist all personnel with training problems.

Education Program

After one year of full-time employment with the Company, employees are entitled to special educational assistance benefits. The Company will pay the full cost of one course (one subject) per term provided that in the Company's sole discretion the course is directly related to the employee's present or planned work assignment. The General Manager has additional information.

Information Dissemination

The Company will maintain a bulletin board near the time clock for posting all documents, posters, or other information required by local, state, or federal regulatory agencies. This bulletin board will be used to post official company policies, grievance actions, regulations, rules, and general information. It will be known as the Official Bulletin Board.

The Company also will maintain an Unofficial Bulletin Board near the Official Bulletin Board for posting miscellaneous employee information. Both bulletin boards will be maintained under the direction of the Office Supervisor.

Grievance Procedures

The grievance procedure is the problem-solving, dispute-settling machinery of the employer-employee agreement. It is the orderly means by which the employee raises and processes a claim alleging some kind of violation.

The procedure outlined below can satisfactorily resolve the overwhelming majority of disputes between parties:

> The employee with the grievance and his immediate supervisor will discuss the problem. If the employee has a legitimate grievance, the supervisor will take the necessary corrective action, or he will forward the matter to a level of management that has the necessary authority to adjudicate properly or resolve the grievance.

The Company prides itself on having a free flow of information between employees at all levels and management. Individual complaints and other work-related problems are usually best handled through informal discussion with supervisors. Every employee should feel free and is encouraged to bring such matters to the attention of his supervisor for quick action. The Company recognizes, however, that some problems cannot be handled satisfactorily in this manner. An employee who feels that he has such a problem should present it to the General Manager. The General Manager will take such action as he deems appropriate, which may include presenting the problem to the management of Four Forty-Four Systems, Incorporated, for full investigation, discussion, and decision.

Causes for Suspension, Demotion, or Dismissal

The following are examples of reasons or causes for a breakdown in effective and harmonious employment and as such are grounds for suspension, demotion, or dismissal from the Company:

1. Incompetence or inefficiency in the performance of assigned duties
2. Insubordination, including but not limited to refusal to do assigned work
3. Carelessness or negligence in the performance of duty or in the care or use of Company property
4. Taking or using Company property without proper authorization
5. Possessing or drinking alcoholic beverages on the job, or reporting to work while intoxicated
6. Possession of, use of, or addiction to narcotics or any controlled dangerous substance
7. Personal conduct that reflects unfavorably on the Company
8. Engaging in political activity during working hours
9. Conviction of any crime involving moral turpitude, larceny, theft, or felony
10. Repeated and unexcused absences or tardiness, such as three tardy arrivals or two unauthorized absences in any pay period
11. Repeated instances of absence for ordinarily excusable or authorizable reasons after sick leave benefits have been exhausted
12. Abuse of employee privileges, including but not limited to sick leave
13. Falsifying any information supplied to the Company, including but not limited to data on application forms, employment records, or any other information required by the Company; subject to enforcement within thirty days after discovery

14. Violation or refusal to obey safety rules or regulations imposed by the Company or any appropriate federal, state, or local government agency
15. Offering anything of value or offering any service in exchange for special treatment in connection with the employee's job or employment, or accepting anything of value or any service in exchange for granting special treatment to a fellow employee or to the public
16. Abandoning one's position, or absence without notification for two or more days
17. Advocacy to overthrow the federal, state, or local government by force, violence, or unlawful means
18. Behavior that has an adverse affect on the work performance of the employee or other employees

Any employee may be suspended immediately if there has been a violation of federal, state, or local laws or company rules and regulations. The suspension may be with or without pay, depending on the individual circumstances. A suspended permanent employee may have access to the General Manager.

Safety

The supervisory personnel within the Company will include safety precautions and standards in training all employees.

The General Manager is assigned the responsibility of making all routine safety inspections within the shop, office, and installation sites. The Installation Supervisor may be designated by the General Manager to make safety inspections of installation sites.

The safety committee will be composed of one representative from each department in addition to the supervisors of each department.

Security

The security of the shop, offices, vehicles, and materials of the Company is the responsibility of all employees. Thefts or vandalism, or attempts of thefts or vandalism should be reported immediately to the General Manager through the employee's immediate supervisor.

An integral part of the security system is the proper control of keys used in the office and shop. Such control must be uniformly applied to all keys used. The following are rules and regulations governing the use of keys:

1. A key issued to an employee of the Company is nontransferrable. It may not be reissued or loaned to another individual.
2. The control of keys and the security of the facility (or facilities) governed by those locks is the direct responsibility of the employee who has been issued the keys.
3. No duplication of keys is allowed. If duplicates are needed, they are to be supplied through the office of the General Manager.
4. Lost keys are to be reported promptly to the General Manager.

Since our business is the installation of alarm systems, we should exercise the utmost care with the system that protects our own building. Persons who have been assigned the responsibility for opening and closing the facilities are charged with the responsibility of testing and setting the alarm system.

Other Policies

Dress. At all times, employees' dress is expected to be appropriate for the type of work they are assigned.

Personnel Records. Employees are required to notify the Office Supervisor as soon as possible of any changes in address, telephone number, legal name, number of dependents, and name, address, and telephone number of next of kin.

Personal Property. The Company is not responsible for the personal property of employees. Employees are asked to limit the amount of personal property on Company premises to a minimum at all times and to use all available precautions to assure safekeeping.

Jury Duty. Jury duty is both a right and a privilege of citizenship. An employee who is called for jury duty must notify his supervisor at once. If he serves for only part of a day, he is expected to work for the balance of the day.

A Final Note

All new employees should be required to read the Organization Manual as soon as possible after employment. After reading the manual, they should sign a statement that they have read and understood its contents.

Whenever revisions are made to the manual, each employee should be made aware of the changes and, after having read and understood the revisions, sign or initial a statement to that effect.

Index